Theory and Technology of Wireless Power Transfer

Shinohara and co-authors present a comprehensive and in-depth discussion of all current wireless power transfer (WPT) methods and meet the growing need for a detailed understanding of the advantages, disadvantages, and applications of each method.

WPT is a game-changing technology, not only for IoT networks and sensors, but also for mobile chargers, long-flying drones, solar-powered satellites, and more, and the list of potential applications will continue to grow. Each author's chapters are based on a minimum of 13 years and a maximum of over 30 years of research experience on selected WPT technologies to explain the theory and advantages and disadvantages of this to various applications. The book provides an insight into WPT theories and technologies, including inductive coupling for short-distance WPT, radio waves for long-distance WPT, optical WPT using lasers, supersonic WPT in water, and more. The characteristics of each WPT method are compared theoretically and technically. The differences of each WPT method are explained with reference to the different theories, techniques, and suitable applications. The reader will gain an understanding of the recent and future commercial market and regulations regarding WPT. They will be able to apply this knowledge to select the appropriate WPT method for their desired application.

This book is appropriate for students, WPT researchers, and engineers in industry who are developing WPT applications.

Theory and Technology of Wireless Power Transfer

Inductive, Radio, Optical, and Supersonic Power Transfer

Naoki Shinohara, Nuno Borges Carvalho,
Takehiro Imura, Tomoyuki Miyamoto,
Kazuhiro Fujimori, and Alessandra Costanzo

CRC Press
Taylor & Francis Group
Boca Raton London New York

CRC Press is an imprint of the
Taylor & Francis Group, an informa business

First edition published 2024
by CRC Press
2385 NW Executive Center Drive, Suite 320, Boca Raton FL 33431

and by CRC Press
4 Park Square, Milton Park, Abingdon, Oxon, OX14 4RN

CRC Press is an imprint of Taylor & Francis Group, LLC

© 2024 [Naoki Shinohara, Nuno Borges Carvalho, Takehiro Imura, Tomoyuki Miyamoto, Kazuhiro Fujimori, and Alessandra Costanzo]

ISBN: 9781032357850 (hbk)
ISBN: 9781032357867 (pbk)
ISBN: 9781003328636 (ebk)

DOI: [10.1201/9781003328636]

Typeset in Minion
by Newgen Publishing UK

Contents

About the Authors

Alessandra Costanzo is full Professor at the University of Bologna, Italy since 2018. She is IEEE Fellow, class 2022, for contribution to "nonlinear electromagnetic co-design of RF and microwave circuits".

Her current research activities are dedicated to the design of entire wireless power transmission systems, for several power levels and operating frequencies. She has developed efficient design procedures based on the combination of electromagnetic and nonlinear numerical techniques, adopting both far-field and near-field solutions, thus creating the bridge between system-level and circuit-level analysis techniques of RF/microwave wireless links. She has accomplished this goal by means of a general-purpose approach combining electromagnetic (EM) theory, EM simulation inside the nonlinear circuit analysis, based on the Harmonic Balance Technique. She is currently the principal investigator of many research and industrial international projects at microwave and millimeter wave dedicated to Industrial IoT, and smart mobility.

She has authored more than 300 scientific publications on peer reviewed international journals and conferences and several chapter books. She owns four international patents. She is past associate editor of the IEEE Transaction on MTT, of the Cambridge Journals IJMWT and JWPT. She is Vice-President for publication of the IEEE CRFID and has been conference chair of EuMC2022 and of IEEE WiSEE 2023. She is co-chair of the Steering Committee of the IEEE WPTCE.

Naoki Shinohara received his PhD (Eng.) degree in electrical engineering from Kyoto University in 1996 and has been a professor there since 2010. He was the first chair of IEICE Wireless Power Transfer and currently chairs the Wireless Power Transfer Consortium for Practical Applications (WiPoT), Wireless Power Management Consortium (WPMc), and URSI

Commission D. He is also Vice Chair of the Space Solar Power Systems Society and is advisor to the Japan Society of Electromagnetic Wave Energy Applications. He is a member, founder, and regional coordinator of several IEEE groups. He is also executive editor of *International Journal of Wireless Power Transfer* and has published several books and articles in both English and Japanese.

Nuno Borges Carvalho is Professor and Senior Research Scientist with the Institute of Telecommunications, University of Aveiro, and an IEEE Fellow. He received his doctoral degree in electronics and telecommunications engineering from the University of Aveiro in 2000. His main research interests include software-defined radio front-ends, wireless power transmission, nonlinear distortion analysis in microwave/wireless circuits and systems, and measurement of nonlinear phenomena. He is the co-inventor of six patents. Dr Borges Carvalho is a member of IEEE MTT ADCOM, belongs to the technical committees MTT-24 and MTT-26, and is Chair of the URSI Commission A (Metrology Group). He is a Distinguished Lecturer for the RFID-Council and was a Distinguished Microwave Lecturer for the IEEE Microwave Theory and Techniques Society.

Takehiro Imura received his DEng degree in electrical engineering from The University of Tokyo, Tokyo, in 2010. In 2019, he joined the Department of Electrical Engineering, Tokyo University of Science, as an Associate Professor. He is currently investigating wireless power transfer using magnetic resonant coupling and electric resonant coupling. His research interests include electric vehicle in-motion connected to renewable energy, sensors, and cancer treatment. He is a member of IEEE, IEICE, and the Society of Automotive Engineers of Japan (JSAE).

Tomoyuki Miyamoto received his PhD (Eng.) in electrical engineering from Tokyo Institute of Technology in 1996. He has worked at Tokyo Institute of Technology since then and is currently an Associate Professor at its Institute of Innovative Research. His current research focuses on optical wireless power transmission (OWPT). He is a member of IEEE Photonics Society, IEICE, the Institute of Electrical Engineers of Japan (IEEJ), the Laser Society of Japan (LSJ), and the Japan Society of Applied Physics (JSAP). He also held the first international conference on OWPT in 2019 and was the Conference Chair from 2019 to 2022.

Kazuhiro Fujimori is Associate Professor of Natural Science and Technology at Okayama University. He received his PhD (Eng.) in electrical and computer engineering from Yokohama National University in 1999. His research addresses small antennas for mobile communications, active integrated antennas, microwave wireless power transfer, and ultrasonic wireless power transfer. He was IEEE AP-S Kansai Joint Chapter vice chair. He is also an executive committee member of IEICE Wireless Power Transfer. Dr Fujimori is also a member of the Wireless Power Transfer Consortium for Practical Applications (WiPoT), the Japan Society of Applied Physics, and the Institute of Electrical Engineers of Japan.

CHAPTER 1

Introduction

ELECTRICITY HAS BECOME AN indispensable part of our lives, with its importance being likened to that of air for humans. In the 19th century, the development of the electrical theory promoted the practical application of electricity. The invention of the voltaic pile (1800), the invention of an early form of an alternating current electrical generator by Pixii (1832), the practical use of generators by Siemens and Gramme (1866), and the invention of the incandescent light bulb by Edison (1882), etc., practically occurred in series with the discovery of the basic theory of Coulomb's law (1785), Ampere's law (1820), Ohm's law (1826), and Faraday's law (1831), etc. It was the most productive period for both basic theory and practical application of electricity. We have relied on two methods of using electricity: battery and wired power transmission from a power station.

Meanwhile, wireless technology with electromagnetic waves emerged in the 19th century. Maxwell's equations, the most crucial theory of electromagnetic waves and fields, were developed in 1864. After Hertz conducted the first trial of electromagnetic wave generation in 1888, Marconi established the Wireless Telegraph & Signal Company in 1897, and successfully achieved the first transatlantic wireless communication in 1901. This technology was established concurrently with the development of practical applications of electricity.

DOI: 10.1201/9781003328636-1

Nikola Tesla, a unique genius, was born in Austria in 1856, the most innovative period for science and engineering. He combined electricity and wireless technology and created the "wireless power transmission or transfer (WPT)" technology at the end of the 19th century. Some WPT field experiments were carried out in the US. He performed an inductive (near-field) WPT experiment in the laboratory and a radiative (far-field) WPT field experiment. Finally, he proposed the "World Wireless System," with which wireless electricity could be transmitted anywhere on the Earth by resonance coupling technology. Unfortunately, his dream ended abruptly as this technology was too far ahead of its time.

In the 1960s, the far-field WPT was back in the spotlight with the new technology of microwave and lasers. The 1960s were a glorious period for the US and the Soviet Union regarding space development. Except for solar cells or batteries, supplying power in space is extremely difficult. However, if we can transmit the power without a wire, it is suitable in space. Therefore, the US considered the far-field WPT for use in space in the 1960s [1–3]. One of the hopeful applications of the far-field WPT in space is a solar power satellite (SPS) [4] (recently termed as space-based solar power [SBSP]). The generated power in SPS (SBSP) in a geostationary orbit will be transmitted via microwave or laser to the ground, and stable and CO_2-free electricity will be obtained. Brown also conducted field experiments of the far-field WPT for flying drones in 1964/68 and of a 1-mile distance WPT in 1975 [2]. The SPS has not been launched in the 21st century; however, some research and development projects are being carried out worldwide. The far-field WPT via microwave has been commercially applied for RF-ID (Radio Frequency Identification) since the 1990s and for multi-purpose wireless chargers or wireless power supplies since 2010 [5–7]. Some startup companies are now also promoting a commercial optical WPT system.

In the 1970s, the near-field WPT was applied to charge an electric vehicle (EV). Prof. Don Otto of the University of Auckland in New Zealand proposed an inductively powered vehicle in 1972, using the power generated, at 10 kHz, by a force-commutated sinusoidal silicon-controlled rectifier inverter [8]. In the 1980s, another EV-WPT project was performed in California, US, called the PATH (Partner for Advanced Transit and Highways) project [9]. Currently, some commercial wireless chargers can be bought for EVs. Additionally, since the 1980s, a low-power near-field wireless charger has been applied for an electric toothbrush, modeless phone, IC-Card (Integrated Circuit Card) for transportation, etc. After the revolution of the near-field WPT named a resonance coupling WPT in 2007 by the Massachusetts Institute of Technology (MIT) [10], the near-field WPT has been generally expanded to charge a smartphone, wireless earphone, etc. [11].

This book discusses all WPT technologies, including near-field WPT by inductive coupling (Chapter 2), near-field WPT by capacitive coupling (Chapter 3), far-field WPT via radio waves (Chapter 4), optical WPT (Chapter 5), and ultrasonic WPT (Chapter 6). Finally, we summarize all WPTs and highlight the pros and cons of each WPT technology (Chapter 7). The reader will learn all about the WPT in this book. WPT technology will soon make Tesla's dream of "unconscious electricity at anytime and anywhere on the Earth" a reality. Only then will electricity be like air, critical for our life but unconscious.

REFERENCES

[1] W. C. Brown, "Adapting Microwave Techniques to Help Solve Future Energy Problems," *1973 G- MTT International Microwave Symposium Digest of Technical Papers*, vol. 73, no. 1, pp. 189–191, 1973.

[2] W. C. Brown, "The History of Power Transmission by Radio Waves," *IEEE Transactions on Microwave Theory and Techniques*, vol. 32, no. 9, pp. 1230–1242, 1984.

[3] W. J. Robinson, Jr., "The Feasibility of Wireless Power Transmission for an Orbiting Astronomical Station," *NASA Technical Memorandum*, X-53701, 1968.

[4] P. E. Glaser, "Power from the Sun," *Science*, vol. 162, no. 3856, pp. 857–886, 1968.

[5] N. Shinohara, "*Wireless Power Transfer via Radiowaves (Wave Series)*," ISTE Ltd. and John Wiley & Sons, Inc., pp. 1, 2014. ISBN 978-1-84821-605-1

[6] N. Shinohara (ed.), "*Recent Wireless Power Transfer Technologies Via Radio Waves*," River Publishers, pp. 5, 2018. ISBN 978-879360-924-2

[7] N. Shinohara and J. Zhou (ed.), "*Far-Field Wireless Power Transfer and Energy Harvesting*," Artech House, pp. 8, 2022. ISBN 978-1630819125

[8] D. Otto, *New Zealand Patent Number*, vol. 167, no. 422, 1974.

[9] J. G. Bolger, F.A. Kirsten and L.S. Ng, "Inductive Power Coupling for an Electric Highway System," *IEEE 28th Vehicular Technology Conference*, vol. 28, pp. 137–144, Mar. 1978

[10] A. Kurs, A. Karalis, R. Moffatt, J. D. Joannopoulos, P. Fisher and M. Soljačić, "Wireless Power Transfer Via Strongly Coupled Magnetic Resonances," *Science Express*, vol. 317, no. 5834, pp. 83–86, 2007.

[11] N. Shinohara (ed.), "*Wireless Power Transfer: Theory, Technology, and Applications*," Inst of Engineering & Technology, pp. 8, 2018. ISBN 978-178561-346-3

Wireless Power Transmission by Magnetic Field

THIS CHAPTER DESCRIBES A method of transmitting power wirelessly by coupling with a magnetic field. The history of wireless charging by electromagnetic induction has a long history, starting around the 1970s, but the research and development that has led to the current situation began in 1991 at the University of Auckland, and has created a major trend [1,2]. The technology of wireless power transfer (WPT) using magnetic field coupling has been commercialized and used in cordless phone handsets, electric toothbrushes, and, industrially, in factory transport systems. In recent years, the increase in the number of mobile devices has increased the demand for easier charging operations, and the Qi standard was established in 2010, which was later adopted by Apple in 2017, and has begun to spread elsewhere. On the other hand, its short transmission/reception distance has limited its use. In the meantime, wireless charging by magnetic field coupling has evolved further, thanks to a major technological breakthrough announced by MIT in 2007 [3], in which magnetic resonance coupling, which makes good use of the resonance phenomenon, enables highly efficient wireless power transmission over an astonishing distance of 2 meters. The imminent commercialization of wireless charging for electric vehicles (EVs) will also spur attention to this technology.

DOI: 10.1201/9781003328636-2

2.1 OVERVIEW OF MAGNETIC COUPLING

2.1.1 Frequency in Terms of Magnetic Field Coupling

Frequency is a very important factor in wireless power transmission; as shown in Figure 2.1, the kHz and MHz bands are mainly used in magnetic field coupling. Similar to the evolution of switching frequencies in power supplies, but starting with the kHz band, the use of the MHz band began mainly after the MIT announcement. From the point of view of the Radio Law, the available frequencies are finite, therefore it is necessary to avoid frequencies used elsewhere. Technically speaking, the technologies used for magnetic field coupling are slightly different between the kHz and MHz bands, as explained in this section.

The situation differs from country to country, but here the current situation in Japan is described, where a ministerial ordinance revision on March 15, 2016, made it possible to apply for type designation for wireless power transmission of magnetic field type: 7.7 kW wireless charging for EVs in the 85 kHz band and 100 W wireless charging for mobile devices and other home appliances in the 6.78 MHz band, both of which are being promoted by the government (Figure 2.1). The 6.78 MHz band is also the ISM band.

Wireless power transmission must inevitably be high frequency in order to achieve high efficiency. As a rule unique to Japan, high-frequency equipment exceeding 10 kHz and 50 W falls under the category of high-frequency use equipment, so it is necessary to apply for a permit. For example, by using large coils and large resonant capacitors, wireless power transmission of high power at commercial frequencies of 50 Hz or 60 Hz is possible. However, this is not common because of the larger volume and

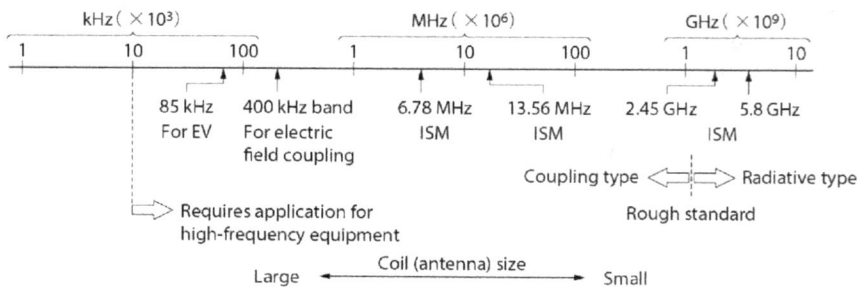

FIGURE 2.1   Frequencies and types of wireless power transfer.

Source: [4].

weight, and it may be caught in harmonics rather than the fundamental frequency at which it is operating. Rather, it is more commonly used in the region below 50 W. For example, at a frequency of about 100 kHz, about 5 W, or about the same power used in USB, is more commonly used for hobby purposes such as simple experiments.

The Qi standard, which began in July 2010, also does not require an application for high-frequency use equipment, because the Qi standard is about 5 W. The frequency used is also 110–205 kHz for the Qi standard. The Qi standard is currently being extended to 15 W. Furthermore, the trend toward higher power is beginning, and the Wireless Power Consortium (WPC) is preparing several higher power standards. Mobile Laptop 300W, Light Electric Vehicle 500W, Cordless Kitchen 2.2kW.

2.1.2 Principle of Magnetic Field Coupling (Electromagnetic Induction, IPT)

Wireless power transfer by magnetic field coupling, also called inductive power transfer (IPT), uses the phenomenon of electromagnetic induction (Figure 2.2). When a voltage is applied to the coil on the power transmission side (primary side), a current flows in the primary side. When current flows, a magnetic flux (magnetic field) is generated. The magnetic flux is chained to the coil on the receiving side (secondary side, pickup). Therefore, this flux is called the main flux. The magnetic flux that is not interlinked and is generated only around its own coil is called the leakage flux.

A voltage is generated on the secondary side in the direction to cancel the magnetic flux chained to the secondary side. This is called induced electromotive force. The induced electromotive force, or voltage, generated on the secondary side acts as a new power source for the coil on the secondary side. After that, current flows in the circuit connected to the secondary coil. Since the current is generated by induced electromotive force, it is also called induced current. This is the basis of wireless power transmission by electromagnetic induction. This principle of operation is independent of frequency. It is not possible with direct current, but with alternating current, the operation is the same whether it is in the kHz or MHz band. Also, the operation so far has nothing to do with resonance.

If the distance between the coils is small and the coupling is strong, this alone can transmit power. However, if the coils are only tightly

Induced current

Induced electromotive force

Receiving coil (secondary side)

Main magnetic flux

Flux leakage

Transmitting coil (primary side)

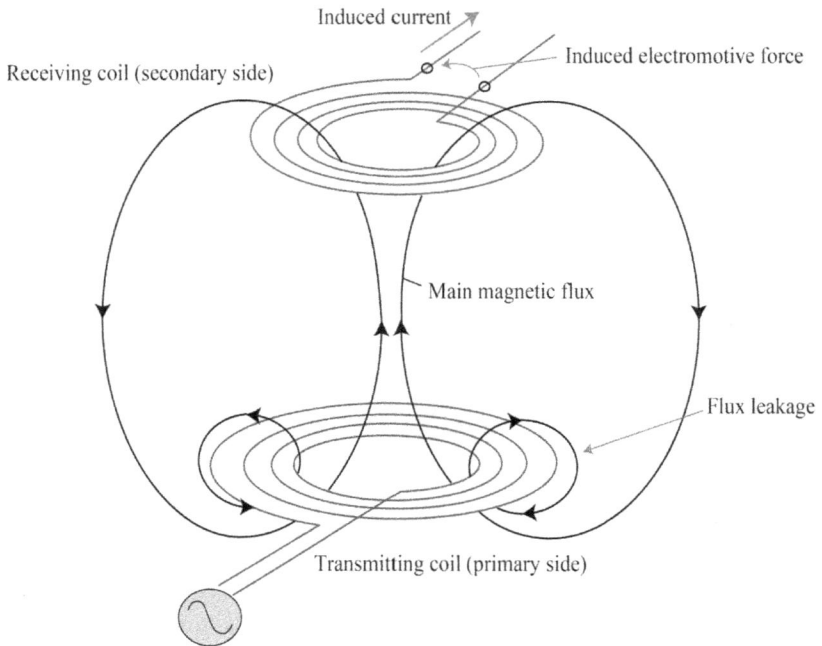

FIGURE 2.2   Principle of electromagnetic induction.

coupled, leakage flux is large and high efficiency and high power cannot be achieved. It is impossible to strengthen the coupling unless a magnetically closed circuit is made of ferrite (Figure 2.3). Such close coupling can no longer be called wireless power transmission. When wireless power transmission is desired, the distance between coils naturally increases. In such a case, the leakage flux becomes dominant and the chain flux decreases.

The coil also has an inductance L. Although this inductance L, which can generate a magnetic field, is necessary for the coupling itself, it is also an obstacle when considering higher efficiency and higher power for wireless power transmission. Therefore, a capacitance C of a capacitor is needed to cancel out this inductance L. In other words, resonance is necessary. By using resonance, high efficiency and high power can be achieved even if the air gap between the transmitting and receiving coils is widened. The specific question then becomes how to use resonance. Resonance requires an explanation of the difference between conventional electromagnetic induction and magnetic field resonance coupling, which will be discussed in more detail in Section 2.1.3.

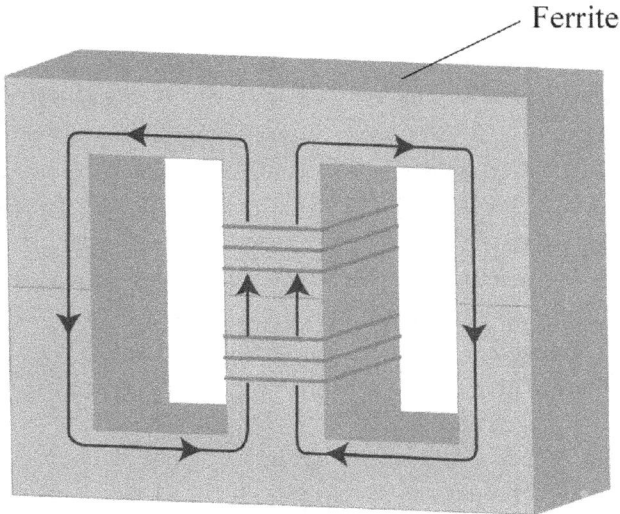

FIGURE 2.3   Ferrite for stronger magnetic coupling.

2.1.3 Principle of Magnetic Field Resonance Coupling

We now discuss the resonant circuit configuration. The difference between conventional electromagnetic induction and the magnetic field res-onant coupling presented by MIT in 2007 has long been discussed. Many attempts have been made to classify them by frequency, by Q-value, by the difference between open and short types, and by the presence or absence of reflection as represented by the S-parameter. In recent years, it has become the consensus of the academic community that it is appropriate to classify by circuit topology. In other words, when comparing conven-tional electromagnetic induction and magnetic field resonant coupling, it is sufficient to explain the resonance phenomenon in terms of circuit top-ology. Therefore, differences in frequency and Q-values are not relevant here. To sum up, magnetic field resonant coupling uses electromagnetic induction as a phenomenon, and the difference between magnetic field res-onant coupling and the conventional method is that the resonant circuit is unique, which enables high efficiency and high power with a large air gap. Of course, the high Q-value and the use of 10 MHz for this purpose were also background factors. In fact, the circuit topology itself had already been presented by the University of Auckland and others around 2000 [5], but the MIT announcement in 2007 was significant because of its insistence on the above-mentioned feature of achieving a large air gap [3].

2.1.4 Effects of Magnetic Field Resonance Coupling

First, we will explain the difference between the effects of conventional electromagnetic induction and magnetic field resonant coupling. Conventional electromagnetic induction is here referred to as non-resonant. It is then denoted by N. The type with no resonant circuit on both the transmitter and receiver sides is called N–N type. The S–S type has resonant circuits on both the transmitter and receiver sides, as in magnetic field resonant coupling, and a resonant capacitor is inserted in series to cause resonance. There are many ways of making resonance, such as the S–P type in which the power receiving side resonates in parallel, but the S–S type is explained here as a representative example. The power supply is assumed to be a general constant voltage source.

Figure 2.4 shows schematic diagrams of the N–N and S–S types. Figure 2.5 shows the equivalent circuits of the N–N and S–S types. The impedance of the load is shown as Z_L; a purely resistive load R_L is often

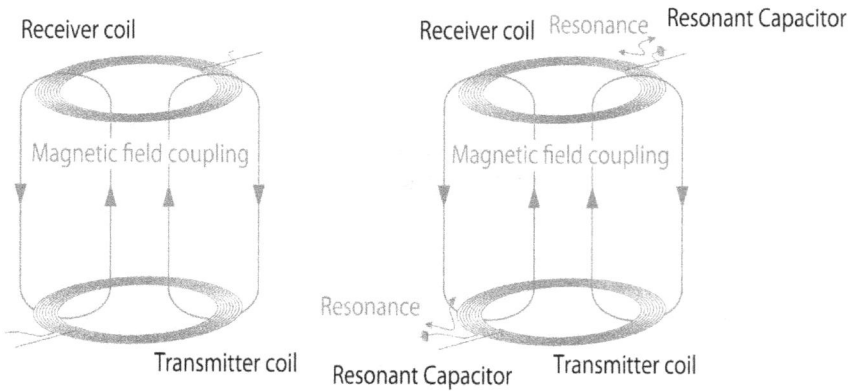

FIGURE 2.4   Schematic diagram of N–N and S–S.

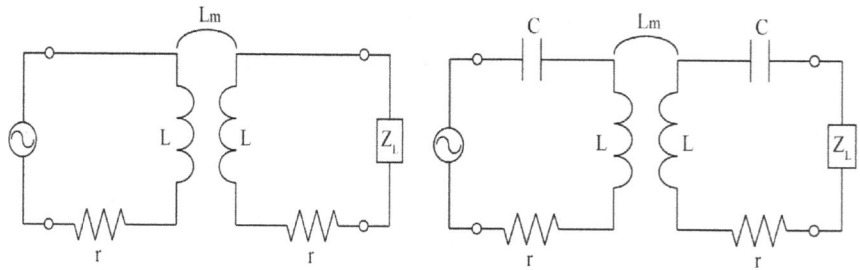

FIGURE 2.5   Equivalent circuit of N–N and S–S.

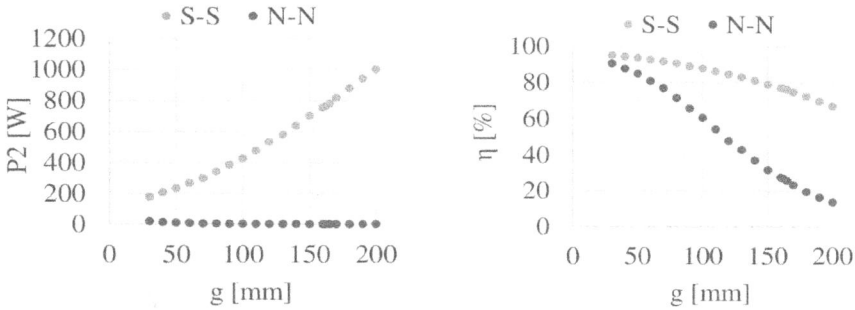

FIGURE 2.6   Comparison of power and efficiency between N–N and S–S.

Source: [4].

assumed since it appears almost purely resistive when rectified in the kHz band. First, power and efficiency versus air gap are shown in Figure 2.6.

As can be seen in Figure 2.6, for all air gaps, S–S achieves higher power and higher efficiency than N–N. This is especially true when the air gap is large. It is important to note that S–S achieves higher power when the air gap is increased. The general feeling is that when the coupling weakens, the power also decreases. In fact, for N–N, the power becomes smaller when the coupling weakens. However, S–S is different. This can be understood by focusing on the impedance seen from the power transmission side. This is related to the gyrator theory and K-inverter theory discussed below, but here is a simple explanation: as the air gap increases, the impedance visible from the transmission side decreases, resulting in higher power. Naturally, when the coupling is completely broken, the received power P_2 also becomes 0 W, but just before that, the power increases as shown in Figure 2.6. The efficiency is as you intuitively expect. As the coupling weakens, the efficiency also decreases. However, it is important to note that the degree of the decrease also depends on the circuit topology.

2.1.5 Seamless Comparison (N–N, N–S, S–N, S–S)

In the previous section, we only compared the simplest N–N and S–S to help you understand the difference between conventional electromagnetic induction and magnetic field resonance coupling, but in this section, we will use four circuits, N–N, N–S, S–N, and S–S, to illustrate the dominance in a more in-depth manner [5].

The T-type equivalent circuit is shown in Figure 2.7. The impedance of the secondary side is Z_{in2}, and the T-shaped part of the coupling part is

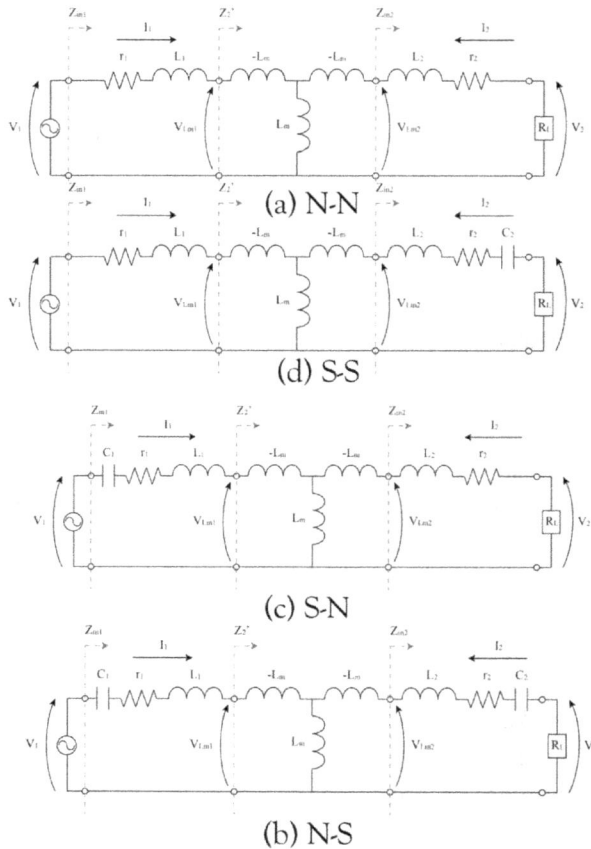

FIGURE 2.7   T-shaped equivalent circuit.

Source: [4].

called K-inverter, or gyrator or immittance characteristic. The impedance (reflected impedance) when looking at the secondary side from the primary side, including the K-inverter, is Z_2'. Finally, let Z_{in1} be the impedance of the primary viewed from the power supply to the load.

First, let us briefly explain the K-inverter theory. The characteristic of this theory is that the numerator and denominator of the impedance of the secondary side are inverted, resulting in the relationship as shown in Equation (2.1). The relationship between large and small loads is reversed. What is interesting is that the $(\omega L_m)^2$ part in the numerator has no imaginary component. In other words, if Z_{in2} is a real number, then this K-inverter theory (gyrator characteristic) shows that the secondary impedance Z_2' from the

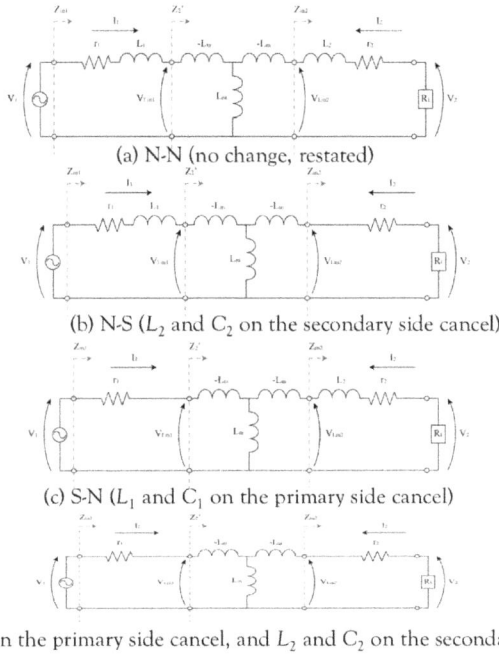

(a) N-N (no change, restated)

(b) N-S (L_2 and C_2 on the secondary side cancel)

(c) S-N (L_1 and C_1 on the primary side cancel)

(d) S-S (L_1 and C_1 on the primary side cancel, and L_2 and C_2 on the secondary side also cancel)

FIGURE 2.8   T-type equivalent circuit (at resonance).

Source: [4].

primary side is also a real number. This is the same for any circuit and is one of the characteristics of the coupling type.

$$Z_2' = \frac{\left(\omega L_m\right)^2}{Z_{in2}} \quad (2.1)$$

Next, we look at the case of a resonant capacitor. The coils on the transmitter or receiver side can cancel each other out. In other words, L can be canceled by C to create a resonant state. The resonant circuit is shown in Figure 2.8. Comparing with Figure 2.7, we can see that L on the transmitting or receiving side is canceled by C. N–N is unchanged, N–S is the secondary side L canceled by C, and S–N is the primary side L canceled by C. S–;S is the transmitting and receiving side Ls on both sides are cancelled out by Cs.

This eliminates L_2 on the secondary side in N–S compared to N–N, resulting in improved efficiency. The degree of improvement is the same as

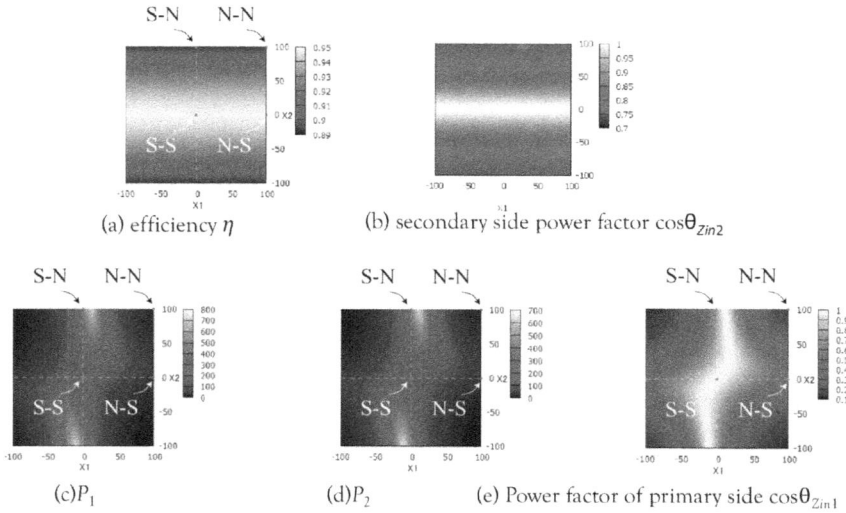

(a) efficiency η

(b) secondary side power factor $\cos\theta_{Zin2}$

(c)P_1

(d)P_2

(e) Power factor of primary side $\cos\theta_{Zin1}$

FIGURE 2.9   Seamless representation of power and efficiency.

Source: [6].

that of S–S. However, the power is lower due to the presence of L_1 on the primary side. S–N, on the other hand, has no L_1 on the primary side, so it produces more power than any of the other four methods. On the other hand, the efficiency remains low because of the presence of L_2 on the secondary side. The efficiency is the same as that of N–N. Therefore, the efficiency is not good. Finally, S–S improves efficiency by eliminating L_2 on the secondary side. This is the maximum efficiency for the circuit topology. Second, the elimination of L_1 on the primary side allows for higher power, not as high as S–N, but still high power. In other words, S–S can achieve high efficiency and high power.

To understand this visually, Figure 2.9 shows a map of the reactance of the primary side as X_1 and the reactance of the secondary side as X_2, changing the capacitance of the capacitor and the values of X_1 and X_2. The values of coils L_1 and L_2 ($L_1 = L_2$) for transmitting and receiving power are fixed; when ωL and $1/\omega C$ are equal, resonance occurs, resulting in cancellation of L and C.

$$X_1 = \left(\omega L_1 - \frac{1}{\omega C_1} \right) \tag{2.2}$$

$$X_2 = \left(\omega L_2 - \frac{1}{\omega C_2} \right) \qquad (2.3)$$

Figure 2.9 shows that the efficiency peaks when the secondary power factor $\cos\theta_{Zin2} = 1$. Also, high power is realized when the power factor of the primary is $\cos\theta_{Zin1} = 1$. The peak efficiency can be reached at S–S and N–S, and high power can be achieved at S–S under these two conditions. The maximum efficiency as a resonance condition of the circuit topology is shown in this figure. However, even in S–S, the maximum efficiency is not achieved just by taking resonance. One condition is missing. That is the condition of optimum load. In other words, the efficiency is ultimately determined by the resistance value of the load. Since the optimum load that can achieve this maximum efficiency depends on the air gap, the load must always be adjusted to achieve the maximum efficiency when the air gap fluctuates.

The equation for the optimum load that results in the maximum efficiency is (2.4), and the equation for the maximum efficiency at that time is (2.5).

$$R_{Lopt} = \sqrt{r_2^2 + \frac{r_2 \left(\omega L_m \right)^2}{r_1}} \qquad (2.4)$$

$$\eta_{max} = \frac{\left(\omega_0 L_m \right)^2}{\left(\sqrt{r_1 r_2} + \sqrt{r_1 r_2 + \left(\omega_0 L_m \right)^2} \right)^2} \qquad (2.5)$$

2.1.6 Coil: kQ Product, Eddy Currents and Proximity Effect

This section discusses coils. The efficiency of a coil is determined by the product of kQ. As shown in Equation (2.6), the coupling coefficient k is the ratio of the mutual inductance L_m to the self-inductance L. And as shown in Equation (2.7), the Q value is a parameter that determines how much energy can be retained. It is the Q value at no load determined by the resonance angular frequency ω, self-inductance L, and internal resistance r.

Eventually, when k and Q are multiplied, self-inductance L disappears, but a large self-inductance L is required to obtain a large mutual inductance

L_m. On the other hand, if many windings are wound to obtain a large L, the internal resistance r will also increase. Therefore, it is necessary to find a balance between L and r.

$$k = \frac{L_m}{L} \tag{2.6}$$

$$Q = \frac{\omega L}{r} \tag{2.7}$$

With direct current, this resistance r is not so large, but at high frequencies, the current flows only on the surface, so the cross-sectional area where the current flows decreases and resistance increases. Figure 2.10 shows the results of analysis of the skin effect at 50 Hz and 85 kHz using the magnetic field analysis software JMAG. In order to suppress this skin effect, a Litz wire is used, which is a copper wire that is finely divided to suppress the skin effect (Figure 2.11).

This Litz wire is made of 301 wires of 0.1 mm each, with 43 bundles wound into seven more bundles to form a single bundle. A measurement of 2.43Ω was obtained using a single wire at 100 kHz, while the Litz wire reduced the current to 1.19 Ω. However, even when Litz wire is used, copper wires are in close proximity to each other, and the magnetic field created by the current flowing in the copper wire affects the adjacent copper wire, causing a bias in the current. This is called the proximity effect (Figure 2.12). The

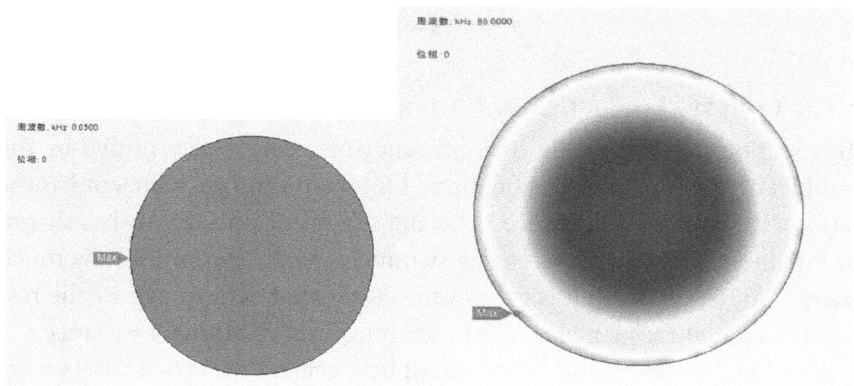

FIGURE 2.10   Skin effect.

FIGURE 2.11   Litz wire.

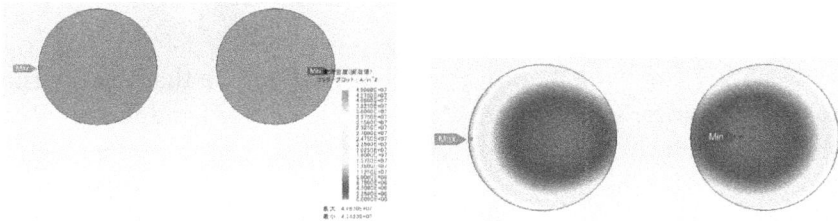

FIGURE 2.12   Proximity effect.

current bias, like the skin effect described earlier, causes a reduction in the cross-sectional area over which the current flows, so even with Litz wire, resistance cannot be lowered to DC resistance.

Due to these phenomena, the value of r is also larger when using high frequency than when using direct current. Therefore, it is not easy to make a coil with a high Q value as targeted. As an example, we will use the coil shown in Figure 2.13, which is 250 mm in length and width, wound with 15 coils of Litz wire, and has a ferrite on the back. The relationship between the Q value, internal resistance r, and self-inductance L of this coil with respect to frequency is shown in Figure 2.14. In Figure 2.14(a), a peak $Q = 525$ is obtained at 186 kHz. In this case,

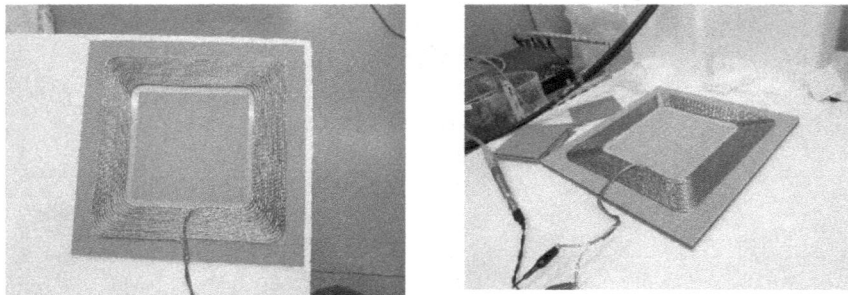

FIGURE 2.13   Coil with ferrite.

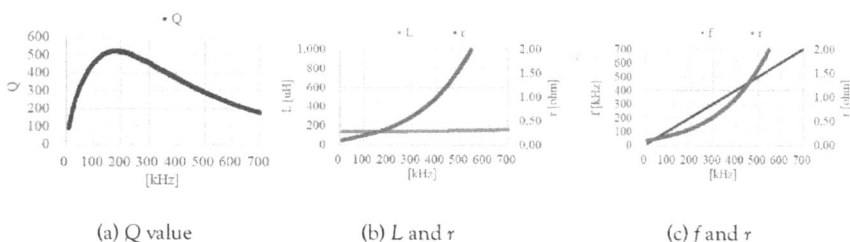

(a) Q value   (b) L and r   (c) f and r

FIGURE 2.14   Q, r, and L.

$L = 139.5\ \mu H$ and $r = 0.3\ \Omega$. In Figure 2.14(b), we see that the internal resistance is increasing exponentially while the self-inductance L is almost constant. Recalling Equation (2.7), if L is constant, the Q value is determined by the relationship between ω and r. Since ω is just the frequency f multiplied by 2π, the graph in Figure 2.14(c) comparing frequency f and internal resistance r shows that the Q value is maximum when r is small relative to f. Therefore, it is important to accurately determine the internal resistance.

On the other hand, it is not easy to accurately determine internal resistance. In recent years, the level of electromagnetic field analysis has greatly improved, and it has become a very effective tool. However, even with electromagnetic field analysis, it is currently very difficult to calculate the resistance increase at high frequencies. Even for single wires, it is extremely difficult, and the analysis of Litz wires, which have a very large number of wires, has not yet reached the level of practical application. On the other hand, methods that use theoretical analysis are gradually approaching the level of practical application, as the time required for analysis is shortening and accuracy is improving [8].

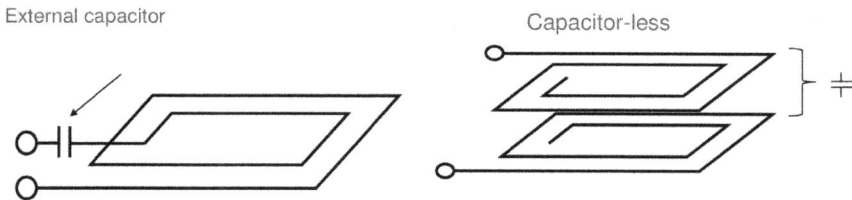

FIGURE 2.15   Short and open type.

2.1.7 Short and Open Types

In the following, kHz-band coils and MHz-band coils are discussed, however, since it is necessary to know about short-type and open-type coils as a prerequisite, they are discussed here [9].

First, there are two types of coils: short type and open type (Figure 2.15). The short type is characterized by the fact that the winding ends back at the feed port. For those unfamiliar with antennas, most people imagine this shape when they think of a typical coil. On the other hand, the open type leaves the end of the entire length of the coil cut off and feeds power from the center of the entire length. The image is that of a half-wavelength dipole antenna wound in a coil shape. Many people familiar with antennas may imagine this type. The short type requires an external capacitor for resonance. On the other hand, the open type can self-resonate using stray capacitance and does not require an external capacitor.

2.1.8 kHz Band Coil

In the case of the kHz band, a short type is generally used in the kHz band, so the explanation is based on the assumption of a short type. A short-type coil cannot resonate by itself, so an external resonance capacitor is required. The use of Litz wire for the coil is often employed as an effective means of reducing the resistance value in order to suppress the skin effect. Power transmission is possible in the kHz band even with an empty core, but ferrites are generally used. Ferrite can be used to increase inductance.

In the case of Figure 2.13, the ferrite is placed on the back of the coil, which increases the coupling from 85.2 µH to 139.5 µH, a 64% increase compared to an air-core coil. Ferrite can also strengthen the coupling. Ferrite is employed for complex reasons other than the above, such as reducing magnetic leakage field. However, attention should be paid to weight. For example, the ferrite in Figure 2.16 measures 100 mm × 100 mm × 5 mm and weighs 250 g. In Figure 2.13, nine ferrites are placed in the back,

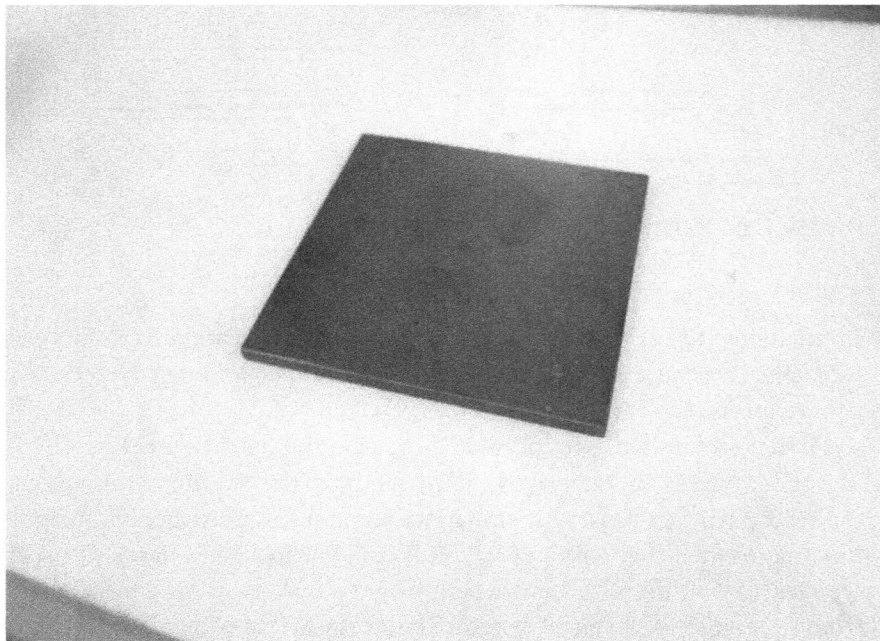

FIGURE 2.16   Ferrite.

so the weight of the ferrites alone is 2.25 kg. When dealing with kW-class, ferrites with a certain thickness are required to avoid magnetic saturation, which makes them heavier. Therefore, in some cases, weight is reduced by thinning out the ferrite. In addition, if there is ferromagnetic material (metal that sticks to the magnet) such as iron on the back, eddy currents will be generated there. Therefore, a non-magnetic (metal that does not stick to the magnet) aluminum plate is often placed behind the ferrite to prevent the magnetic flux from spreading around.

Figure 2.17 shows an image of the completed coil, ferrite, and aluminum plate together. For reference, Figure 2.18 shows the power transmission efficiency when the coils in Figure 2.13 are separated by 222 mm and the coupling coefficient $k = 0.06$. At all frequencies, the load is adjusted to the optimum load that can achieve the maximum efficiency at that frequency. Since the mutual inductance is almost constant with frequency, the peak Q value and the frequency at which the maximum efficiency peaks coincide. In other words, since the maximum efficiency is achieved at the frequency that brings out the best performance of the coil, it can be said that the theoretical maximum efficiency can be achieved for this coil.

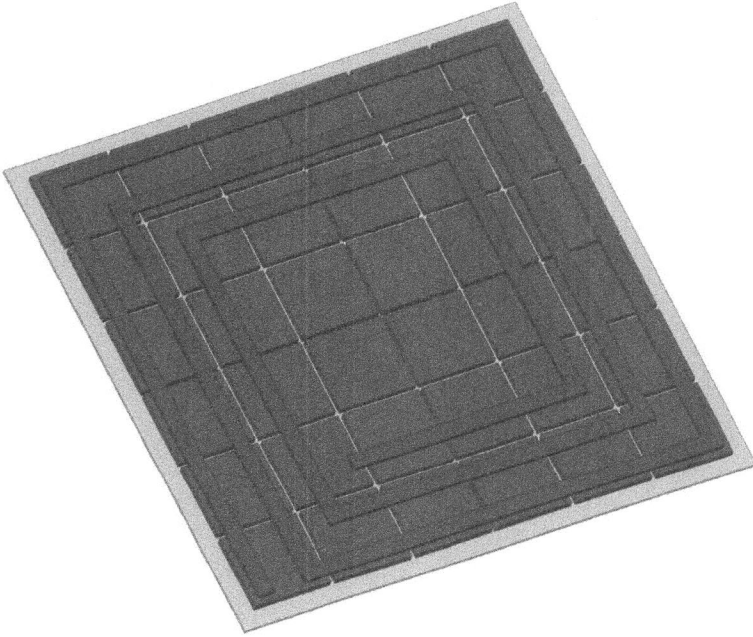

FIGURE 2.17   Coil, ferrite, and aluminum plate.

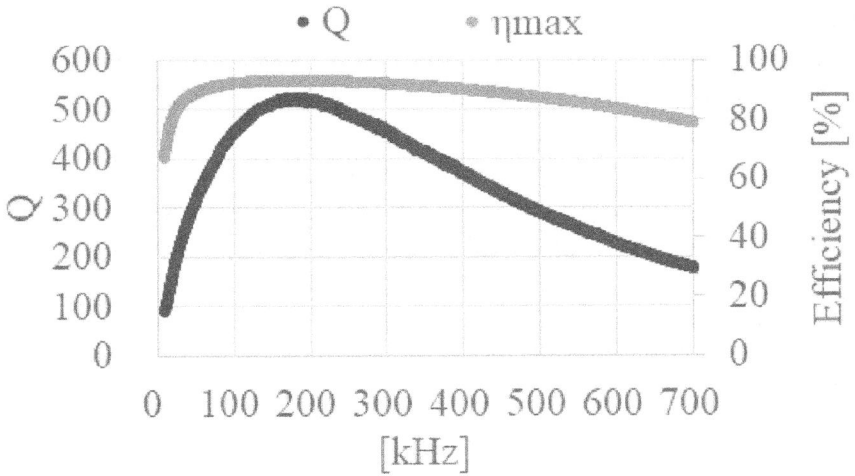

FIGURE 2.18   Efficiency and Q.

From the point of view of resonance, a large L or large C is needed because the resonance is made with an inductance L and a capacitor C, for example, to 85 kHz. The relationship between frequency and angular frequency is shown in Equation (2.8), and the relationship between L and C is shown in Equation (2.9).

$$\omega = 2\pi f \tag{2.8}$$

$$\omega = \frac{1}{\sqrt{LC}} \tag{2.9}$$

Therefore, there is considerable flexibility in how L and C are selected. The size of the coil itself is also quite flexible. Short-type coils use an external capacitor. Since the value of L varies depending on the number of turns and size of the coil, and the external C can be freely selected, there is quite a bit of freedom in design even if the frequency is fixed. For example, a method to increase the mutual inductance L_m is used. In this case, to achieve a large L, the coil is made as large as the application allows, the number of turns is increased, and the ferrite is loaded. In the case of the coil shown in Figure 2.13, $L = 138.5\ \mu H$ at 85 kHz, so $C = 25.3$ nF. For resonance capacitors, laminated ceramic capacitors or film capacitors are used. Due to their small volume, we often see multilayer ceramic capacitors used in recent years.

The worst case can be easily considered from the Q value and input voltage. In a resonant state, the voltage across the capacitor is equal to the voltage across the coil. Especially in the case of magnetic field resonant coupling, a more severe design is required because of the use of a coil with a high Q value. In the case of a 100 V input, a voltage of 10 kV will be applied to the capacitor if the Q value is 100. Although such a worst-case scenario is rare, a coil with a high Q value can provide highly efficient power transmission with a large air gap, but there are many items to be considered when designing the coil. Other considerations, such as the long time required for transients to settle down and transition to a steady state due to the high Q value, must also be taken into account.

2.1.9 MHz Band Coil

Second, in the MHz band, Litz wire cannot be used because the proximity effect is too strong, so single wires are generally used. In addition, ferrite is generally used as a noise countermeasure, which turns the magnetic field into heat and causes loss. Ferrite, which can be used to strengthen coupling and increase inductance as used in the kHz band, is rarely used for general applications, so it must be carefully selected for use. For this reason, most coils in the MHz band are air-core coils. Since they are air-core, the coil part is lightweight and low-cost.

In addition, for the MHz band, both open-type coils and short-type coils can be selected for resonance. With an open coil, resonance can be obtained with floating capacitance, so an external capacitor is not necessary. However, the stray capacitance is a capacitor created between the wires, so it can easily change even if the coil shape changes slightly. On the other hand, a short-type coil cannot resonate by itself as well as in the kHz band, so an external capacitor is required. In the MHz band, the equivalent series resistance (ESR) of external capacitors tends to be high, so care must be taken when selecting capacitors. Also, excellent withstand voltage performance is required. However, since this is a MHz band, even if the inductance L of the coil is small, the capacitance C of the capacitor is small enough to resonate. The quantity of capacitors can be reduced, and they are inexpensive and take up little space.

Figure 2.19 shows an experimental scene of wireless power transmission at 13.56 MHz by connecting a capacitor in series with a short-type coil and making it resonate. As shown, the coil is a single winding, so it is very lightweight. The inductance of the coil is 1.05 μH. The box containing the capacitor is large in the photo because it is for experimental use, but it can be made smaller. The capacitor itself is 138 pF, so it is very small. In Japan, there is an IC card for transportation, Felica (Suica), as an IC card in the MHz band [10]. It uses resonance with a capacitor at 13.56 MHz. The power transmission coil feeds power wirelessly to your card.

In the MHz band, as in the kHz band, it is possible to make a large coil or a small coil at the same frequency, depending on the capacitor and the winding method. Felica is about 8.5 mm × 5.4 mm in length and width, while the coil in Figure 2.19 is 300 mm in diameter. Figure 2.20 shows an open-type coil, but the diameter of the transmitting coil is 960 mm and that of the receiving coil is 380 mm, which also resonates at 13.56 MHz. As shown above, even if the frequency is fixed, the coil size can be selected within a certain range.

2.1.10 Multiband Coils in kHz and MHz Bands

In fact, the only method that has received attention in wireless power transmission has been the use of first-order resonance. However, if one

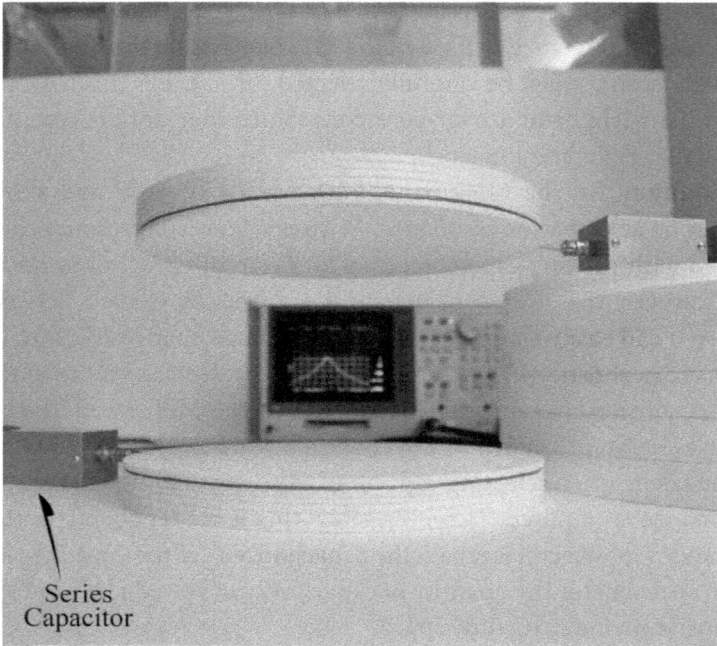

Series
Capacitor

FIGURE 2.19   Experimental view of MHz coil.

utilizes up to second-order resonance, two frequencies can be used with a single coil. This section describes coils that can be used at 85 kHz and 6.78 MHz [11].

When a capacitor is inserted in series with a short-type coil, a resonance occurs. This is referred to here as the first-order resonance. Above that frequency, there is an anti-resonance, and another resonance appears at a frequency above that. This is referred to here as the second-order resonance. By designing these two resonance points well, power transmission is performed at two frequencies. Specifically, the second-order resonance point is created from the coil shape. Since the value of inductance L is determined by the coil wire length, the resonant frequency can be determined by adjusting the distance and length between the coil wires. The next step is to create the primary resonance by inserting an external capacitor.

The power transmission coil and reactance characteristics created by this procedure are shown in Figure 2.21. In this case, as shown in Figure 2.22, the coil on the receiving side for wireless power transmission is either 85 kHz or 6.78 MHz, depending on the intended use. Figure 2.23 shows the power and efficiency of power transmission. Depending on the standard,

FIGURE 2.20   Experimental view of asymmetric coil.

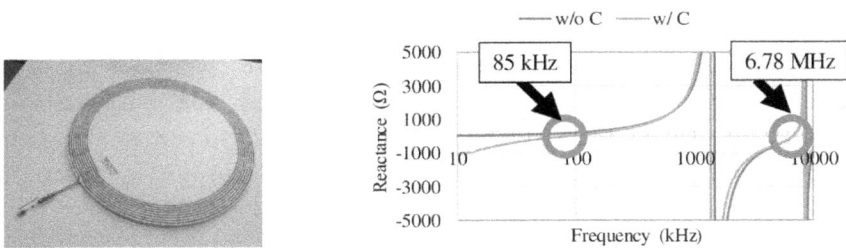

FIGURE 2.21   Dual-frequency shared coil.

there may be a 100 kHz band and a 6.78 MHz band for mobile devices such as cellular phones, and in such cases, a single power transmission coil can be used for both.

2.2 RESONANT CIRCUIT TOPOLOGY OF MAGNETIC FIELD COUPLING

In Section 2.1.5, we discussed the difference between conventional electromagnetic induction and magnetic field resonant coupling by comparing

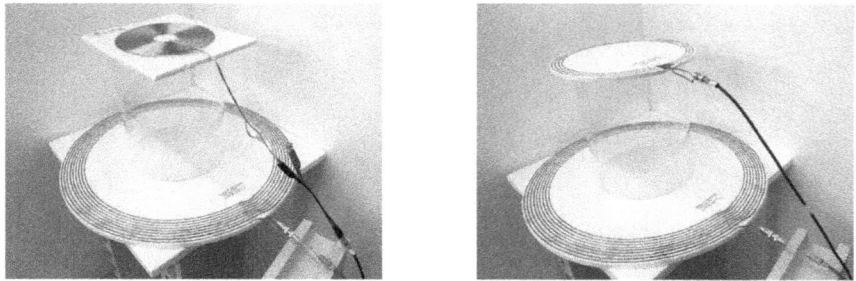

(a) 85kHz   (b) 6.78MHz

FIGURE 2.22   Power transmission using two-frequency shard coils.

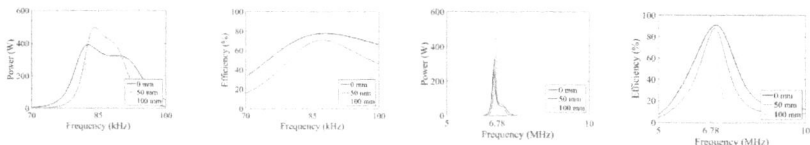

(a) 85kHz, power   (b) 85kHz, efficiency   (c) 6.78MHz, power   (d) 6.78MHz, efficiency

FIGURE 2.23   Efficiency and power of power transmission using two-frequency shared coils.

four types of circuits: N–N, N–S, S–N, and S–S. The S–S circuit, which has one resonant capacitor each on the transmitter and receiver sides, has been called a magnetic field resonant coupling, but if both sides have resonant circuits, the same characteristics as a magnetic field resonant coupling can be produced. Therefore, from this section, we will discuss an advanced version of the magnetic field resonant coupling.

2.2.1 S–S, S–P, P–S, P–P

First, we will discuss S–S, S–P, P–S, and P–P, each of which has one resonant capacitor on the transmitter and receiver sides, remembering that S stands for Series in series and P for Parallel in parallel. Next, we will discuss the power supply, which is generally designed as either a constant-voltage source or a constant-current source, but there are other design methods, such as using a constant current in the transmission coil instead of a power supply. In this section, we will mainly discuss the case of a general constant-voltage source. As a supplement, we will also discuss the characteristics when a constant current source is partially connected.

It is important to note that the constant voltage and constant current characteristics here are for load variations, not for gap variations. When constant-voltage and constant-current characteristics on the load side are described in wireless power transmission, the above assumptions may be made unless otherwise noted. Also, internal resistance is ignored to simplify calculations. Although it cannot be ignored when discussing efficiency, it can be ignored to capture trends when examining constant-voltage and constant-current characteristics. In reality, the tiny internal resistance will affect and deviate slightly from the constant-voltage and constant-current characteristics, but it can still be captured as a general trend.

First, S–S and S–P. S–S and S–P have the same maximum efficiency (Figure 2.24). However, the optimum load at which maximum efficiency can be achieved is small for S–S and large for S–P, so it is necessary to consider this for each application. In addition, S–S and S–P can realize high power. Hence, they are treated as magnetic field resonance coupling.

S–P requires resonance considering the coupling coefficient k to operate as a constant-voltage characteristic; it may be easier to understand S–P as essentially a constant-current characteristic. As a supplement to the constant current source, as shown in Equation (2.11) derived from Equation (2.10), even if only the LC resonance is considered, if a constant current source is connected, the constant current characteristic will appear in a

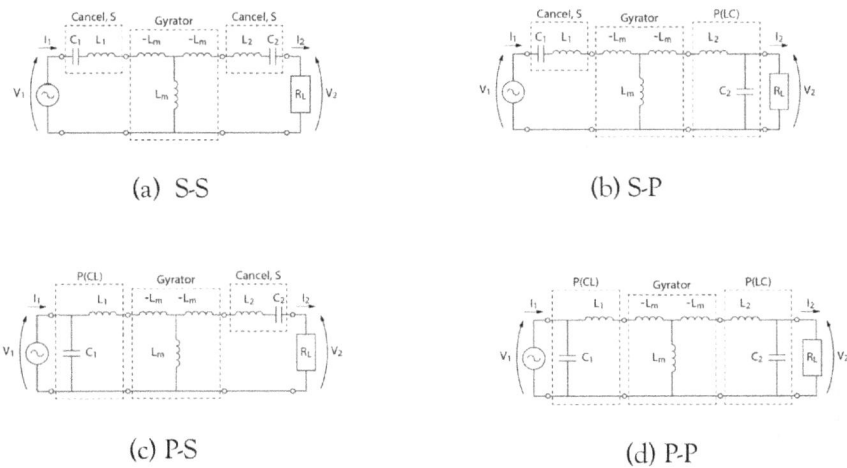

(a) S-S

(b) S-P

(c) P-S

(d) P-P

FIGURE 2.24   S–S, S–P, and P–P equivalent circuits.

clean form on the load side. Thus, C (P and *LC*) in parallel has characteristics like quasi-gyrator characteristics.

$$\begin{bmatrix} V_1 \\ I_1 \end{bmatrix} = \begin{bmatrix} \dfrac{L_m}{L_2} & -j\omega L_m \\ 0 & \dfrac{L_2}{L_m} \end{bmatrix} \begin{bmatrix} V_2 \\ I_2 \end{bmatrix} \tag{2.10}$$

$$I_1 = \frac{L_2}{L_m} I_2 \Leftrightarrow I_2 = \frac{L_m}{L_2} I_1 \tag{2.11}$$

There is also P–S and P–P. In that case, the maximum efficiency of both P–S and P–P is still the same as that of S–S and S–P [12]. However, the power efficiency will be much lower than that of S–S because of the imaginary numbers that remain when designing with normal *LC* resonance. When an additional *L* is added to P–S on the transmission side, the configuration is called LCL-S [13–14]. Some people call it LCL-LC. From this point on, things get increasingly complicated. The reason it is complicated is that it is difficult to explain in a systematic manner because of the different naming conventions, and different circuit designers have different resonance methods and policies. As for naming, we have used S for resonant capacitors in series and P for resonant capacitors in parallel, but with the addition of *L*, it is no longer possible to express the capacitor using only S and P, which were described with the capacitor in mind. On the other hand, using *L* and *C* lacks information on whether it is series or parallel, which is not clear without reading the description. From experience, when we say LCL, we can imagine that *C* is parallel. However, when it comes to LCC, we do not know where series and parallel are until we look at the circuit diagram. Moreover, LCC does not include the *L* in the name of the power transmission coil, so the schematic shows LCCL.

In addition, in the case of the S–P mentioned earlier, we indicated that there is a design method that is aware of constant-voltage characteristics, but even for the same S–P, if the design is aware of LC resonance, an imaginary component will remain on the secondary side and maximum efficiency cannot be achieved [15].

Thus, as the number of circuit points increases, it becomes more difficult to decipher the intent. Another reason is that in wireless power transmission, harmonics are a problem, so L and C may be added as filters to bring the waveform closer to a sine wave, and the purpose of the circuit components is not always clearly specified.

Returning to P–S, P–S is as in Equation (2.12), which does not give a clean gyrator characteristic at P. Therefore, considering only the LC resonance, a detailed analysis is necessary to determine if a constant voltage is produced at the load side when a constant-voltage source is used. Therefore, considering only LC resonance, a detailed analysis is necessary to determine whether a constant voltage is produced on the load side when a constant-voltage source is used. On the other hand, when a constant-current source is used, the load side will have a constant-current characteristic as shown in Equation (2.13).

$$\begin{bmatrix} V_1 \\ I_1 \end{bmatrix} = \begin{bmatrix} \dfrac{L_1}{L_m} & -j\omega L_m \\ 0 & \dfrac{L_m}{L_1} \end{bmatrix} \begin{bmatrix} V_2 \\ I_2 \end{bmatrix} \tag{2.12}$$

$$I_1 = \frac{L_m}{L_1} I_2 \Leftrightarrow I_2 = \frac{L_1}{L_m} I_1 \tag{2.13}$$

Next is P–P. Considering the gyrator characteristics, neither constant-voltage nor constant-current characteristics can be shown when only LC resonance is considered. On the other hand, if we add the assumption that the current flowing in the transmission side coil is constant, the same conditions as for S–P are obtained. Then, the same conclusion is obtained with S–P using a constant current source, even if only LC resonance is considered, the constant-current characteristic will appear on the load side in a clean form. In fact, the University of Auckland's system sometimes uses the condition that the current flowing in the coil on the transmission side is controlled to be constant. In this case, the system can be constructed using characteristics that cannot be seen only in the circuit topology of S–P and P–P.

Table 2.1 shows a summary of the S–S, S–P, P–S, and P–P methods we have looked at so far. Since this method considers only LC resonance, it does not include those that achieve constant-voltage (CV) or constant-current (CC) characteristics using other conditions described in the text.

TABLE 2.1   List of CV and CC Properties of S–S, S–P, P–S, and P–P

	S–S	S–P	P–S	P–P
Constant voltage source	CC	–	–	–
Constant current source	CV	CC	–	–

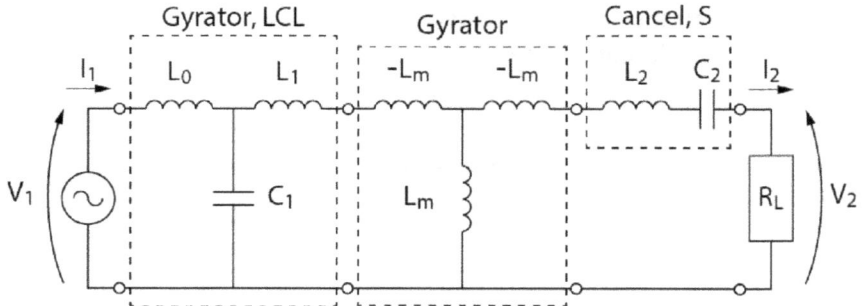

FIGURE 2.25   LCL-S.

Source: [4].

2.2.2 LCL and LCC, etc.

This section describes circuits that are not like S–S, S–P, P–S, and P–P, which contain only one capacitor each on the transmitter and receiver sides, but are slightly developed from these circuits. Although LCL-S was mentioned in part in the previous section, it is explained here collectively. Specifically, LCL-S, LCL-P, S-LCL, P-LCL, LCL-LCL, and LCC. As a general rule, increasing the number of passive components increases losses and costs, so these circuits should be used when there are more benefits to be gained, so it is necessary to pay attention to this area in reading the circuit.

There are many ways to look at LCL, but as the most representative example, I will discuss whether the characteristics that appear on the load side when various circuits are connected to a constant-voltage source are constant-voltage characteristics or constant-current characteristics. When a constant-voltage source was used, S–S had a constant-current characteristic and S–P had a constant-voltage characteristic. Based on this, we will discuss LCL circuits. Also, with the exception of LCC, $L_0 = L_1 = L_2 = L'_0$.

Since the LCL is configured as a gyrator, it can convert an ideal constant-voltage to constant-current characteristic or constant-current to constant-voltage characteristic. Therefore, in a configuration like the LCL-S in Figure 2.25, the load will have a constant-voltage characteristic. In this step-by-step explanation, the constant-current characteristic becomes a constant-current characteristic once at the LCL part, and then the coupling part of the T-shape part is also an ideal gyrator, so the constant-current characteristic becomes a constant-voltage characteristic. At the last S, the characteristics do not change, so the constant-voltage characteristics become constant-voltage characteristics.

In other words, since the gyrator has passed through the gyrator twice, the constant-voltage characteristic is restored. Equation (2.14) shows the gyrator characteristics for two stages, resulting in Equation (2.15). The constant voltage characteristic can be shown as in Equation (2.16).

$$\begin{bmatrix} V_1 \\ I_1 \end{bmatrix} = \begin{bmatrix} 0 & j\omega L_1 \\ j\omega C_1 & 0 \end{bmatrix} \begin{bmatrix} 0 & -j\omega L_m \\ \dfrac{1}{j\omega L_m} & 0 \end{bmatrix} \begin{bmatrix} V_2 \\ I_2 \end{bmatrix} \tag{2.14}$$

$$\begin{bmatrix} V_1 \\ I_1 \end{bmatrix} = \begin{bmatrix} \dfrac{L_1}{L_m} & 0 \\ 0 & \dfrac{L_m}{L_1} \end{bmatrix} \begin{bmatrix} V_2 \\ I_2 \end{bmatrix} \tag{2.15}$$

$$V_1 = \frac{L_1}{L_m} V_2 \Leftrightarrow V_2 = \frac{L_m}{L_1} V_1 \tag{2.16}$$

Next, LCL-P, shown in Figure 2.26, has a constant-current characteristic when a constant-voltage source is used. The explanation is almost the same as for LCL-S, but at the last point P, the constant voltage changes to constant current. As shown in Equation (2.17), there are two gyrator characteristics and, at P, a quasi-gyrator characteristic, so there are three times. The result is not a beautiful gyrator characteristic, but the constant-current characteristic appears as in Equations (2.18) and (2.19).

$$\begin{bmatrix} V_1 \\ I_1 \end{bmatrix} = \begin{bmatrix} 0 & j\omega L_1 \\ j\omega C_1 & 0 \end{bmatrix} \begin{bmatrix} 0 & -j\omega L_m \\ \dfrac{1}{j\omega L_m} & 0 \end{bmatrix} \begin{bmatrix} 0 & j\omega L_2 \\ j\omega C_2 & 1 \end{bmatrix} \begin{bmatrix} V_2 \\ I_2 \end{bmatrix} \tag{2.17}$$

$$
\begin{bmatrix} V_1 \\ I_1 \end{bmatrix} = \begin{bmatrix} 0 & j\omega\dfrac{L_1L_2}{L_m} \\ j\omega C_2\dfrac{L_m}{L_1} & \dfrac{L_m}{L_1} \end{bmatrix} \begin{bmatrix} V_2 \\ I_2 \end{bmatrix}
\tag{2.18}
$$

$$
V_1 = j\omega\frac{L_1L_2}{L_m}I_L \Leftrightarrow I_L = \frac{L_m}{j\omega L_1 L_2}V_1
\tag{2.19}
$$

There is also the S-LCL shown in Figure 2.27. In this case, the first stage is a constant-voltage characteristic, since the initial S has no effect; after the

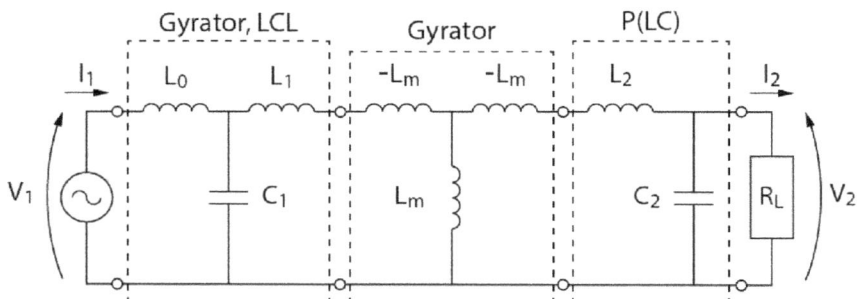

FIGURE 2.26 LCL-P.

Source: [4].

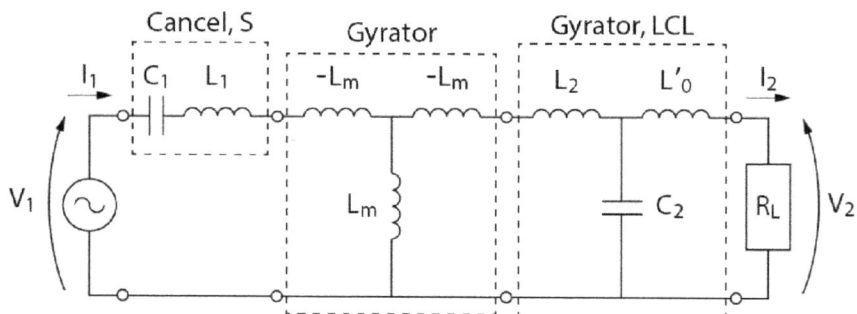

FIGURE 2.27 S-LCL.

Source: [4].

T-shaped equivalent circuit has a constant-current characteristic, the LCL has a constant-voltage characteristic. Equation (2.20) shows the gyrator characteristics for two stages, resulting in Equation (2.21). The constant voltage characteristic can be shown as in Equation (2.22).

$$\begin{bmatrix} V_1 \\ I_1 \end{bmatrix} = \begin{bmatrix} 0 & -j\omega L_m \\ \dfrac{1}{j\omega L_m} & 0 \end{bmatrix} \begin{bmatrix} 0 & j\omega L_0' \\ j\omega C_2 & 0 \end{bmatrix} \begin{bmatrix} V_L \\ I_L \end{bmatrix} \tag{2.20}$$

$$\begin{bmatrix} V_1 \\ I_1 \end{bmatrix} = \begin{bmatrix} \dfrac{L_m}{L_2} & 0 \\ 0 & \dfrac{L_0'}{L_m} \end{bmatrix} \begin{bmatrix} V_L \\ I_L \end{bmatrix} \tag{2.21}$$

$$V_1 = \frac{L_m}{L_2} V_L \Leftrightarrow V_L = \frac{L_2}{L_m} V_1 \tag{2.22}$$

The P-LCL shown in Figure 2.28 is often not taken up because it is tricky. After all, for the same reason as P–S, the beginning P deviates from the gyrator characteristic. Therefore, if we go straight to the LC resonance, the characteristic that appears on the load side when a constant voltage source

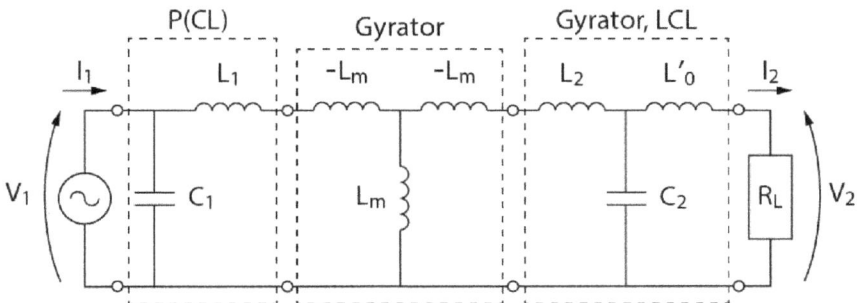

FIGURE 2.28   P-LCL.

Source: [4].

is used will not get to the constant current. On the other hand, if we discuss the case of a constant-current source, this P-LCL has the characteristic that the constant-voltage characteristic is maintained by passing from the constant-current source through three places: P, T-shaped part, and LCL. Equations (2.23)–(2.25) are the mathematical expressions.

$$
\begin{bmatrix} V_1 \\ I_1 \end{bmatrix} = \begin{bmatrix} 1 & j\omega L_1 \\ j\omega C_1 & 0 \end{bmatrix} \begin{bmatrix} 0 & -j\omega L_m \\ \dfrac{1}{j\omega L_m} & 0 \end{bmatrix}
$$

$$
\begin{bmatrix} 0 & j\omega L_0' \\ j\omega C_2 & 0 \end{bmatrix} \begin{bmatrix} V_2 \\ I_2 \end{bmatrix}
\tag{2.23}
$$

$$
\begin{bmatrix} V_1 \\ I_1 \end{bmatrix} = \begin{bmatrix} \dfrac{L_m}{L_2} & j\omega\dfrac{L_1^2}{L_m} \\ j\omega C_1 \dfrac{L_m}{L_1} & 0 \end{bmatrix} \begin{bmatrix} V_2 \\ I_2 \end{bmatrix}
\tag{2.24}
$$

$$
I_1 = j\omega C_1 \frac{L_m}{L_1} V_2 \Leftrightarrow V_2 = \frac{L_1}{j\omega C_1 L_m} I_1
\tag{2.25}
$$

Figure 2.29 shows the LCL-LCL, which consists of an ideal three-stage gyrator: the first LCL provides a constant current, the T-type equivalent

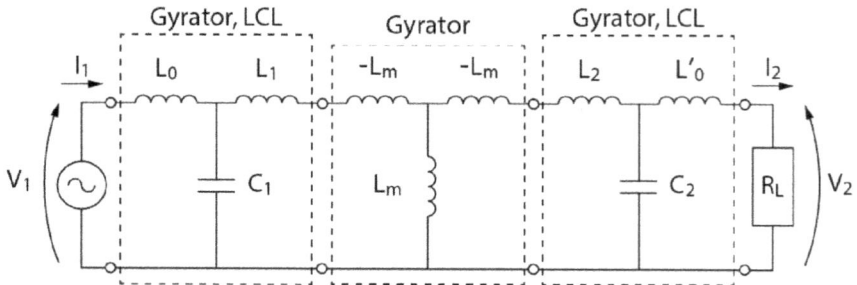

FIGURE 2.29   LCL-LCL.

Source: [4].

circuit provides a constant voltage, and the last LCL provides a constant current. The P part of the P-LCL is firmly in the LCL, so we have reached the constant-current characteristic. Equations (2.26)–(2.28) can be used as formulas. Also, if a constant-current source is connected, a constant-voltage characteristic will appear.

$$\begin{bmatrix} V_1 \\ I_1 \end{bmatrix} = \begin{bmatrix} 0 & j\omega L_0 \\ j\omega C_1 & 0 \end{bmatrix} \begin{bmatrix} 0 & -j\omega L_m \\ \dfrac{1}{j\omega L_m} & 0 \end{bmatrix}$$

$$\begin{bmatrix} 0 & j\omega L_0' \\ j\omega C_2 & 0 \end{bmatrix} \begin{bmatrix} V_2 \\ I_2 \end{bmatrix} \tag{2.26}$$

$$\begin{bmatrix} V_1 \\ I_1 \end{bmatrix} = \begin{bmatrix} 0 & j\omega \dfrac{L_1^2}{L_m} \\ j\omega C_1 \dfrac{L_m}{L_1} & 0 \end{bmatrix} \begin{bmatrix} V_2 \\ I_2 \end{bmatrix} \tag{2.27}$$

$$V_1 = j\omega \frac{L_1^2}{L_m} I_2 \Leftrightarrow I_2 = \frac{L_m}{j\omega L_1^2} V_1 \tag{2.28}$$

The above is a brief explanation, but the same can be considered with a constant-current source for a power supply if it has a clean gyrator characteristic. For example, as mentioned above, if an LCL-LCL is connected to a constant current source, the first LCL provides a constant voltage, the T-type equivalent circuit provides a constant current, and the last LCL provides a constant voltage.

There is also a circuit called LCC [16]–[17]. Looking at the circuit, the LCC is connected symmetrically on the transmitting and receiving sides, so it is sometimes called Double LCC. Also, the L of the transmitting and receiving coils is not included in the name, so the circuit is LCCL-LCCL (Figure 2.30).

The LCC is a derivative of the LCL, and the insertion of C_{1S} has the effect of reducing the L_0 in the LCL and increasing the power [16]. As for the equation that constitutes the LCC, the general framework is the same

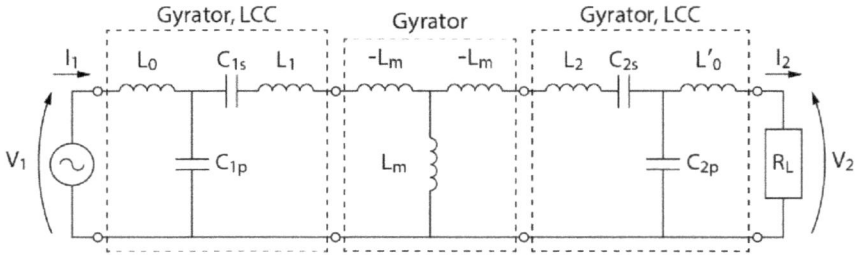

FIGURE 2.30 Double LCC circuit.

Source: [4].

design method as for the LCL that constitutes the three-stage gyrator, which is Equation (2.29). In this case, the resonance conditions are similar. Make the resonances at the three locations of the gyrator's T-type circuit in the secondary circuit the same. Therefore, there are resonances at L'_0 and C_{2P} and resonances at C_{2P} and "L_2 and C_{2S} in series".

Therefore, the resonance condition is Equation (2.30). The primary side can be considered in the same way. The constant-voltage and constant-current characteristics are the same as for LCL, so they are omitted.

$$
\begin{bmatrix} V_1 \\ I_1 \end{bmatrix} = \begin{bmatrix} 0 & j\omega L_0 \\ j\omega C_{1P} & 0 \end{bmatrix} \begin{bmatrix} 0 & -j\omega L_m \\ \dfrac{1}{j\omega L_m} & 0 \end{bmatrix}
$$

$$
\begin{bmatrix} 0 & j\omega L'_0 \\ j\omega C_{2P} & 0 \end{bmatrix} \begin{bmatrix} V_2 \\ I_2 \end{bmatrix}
$$

(2.29)

$$
\omega_0 = \frac{1}{\sqrt{L'_0 C_{2P}}} = \frac{1}{\sqrt{L_2 \dfrac{C_{2S} C_{2P}}{C_{2S} + C_{2P}}}}
$$

(2.30)

Table 2.2 shows a summary of the LCL and LCC we have looked at so far. In this section, the gyrator characteristics are considered and only LC resonance is considered in this method.

TABLE 2.2   List of LCL-related CV and CC Properties

	LCL-S	LCL-P	S-LCL	P-LCL	LCL-LCL	Double LCC
Constant voltage source	CV	CC	CV	–	CC	CC
Constant current source	CC	–	CC	CV	CV	CV

2.2.3 Control Using Circuit Topology Characteristics

In the previous sections, we specialized in the coil area, including circuit topology. From this point on, the system will be explained. From the explanations that have been given so far, you will have understood that efficiency and power can be large or small depending on conditions. Therefore, it is necessary to always achieve high efficiency and to be able to receive the desired and necessary power at the receiving end, even if conditions change.

This power received is referred to here as the desired power. In other words, it is important for a system to achieve high efficiency and desired power. The desired power is not so critical if the load is a resistor or a constant voltage like a battery, but if the load is a constant power load like a motor or electronic equipment, the buffer to store the power is small, so more precise control is required. Therefore, it is necessary to go one step further from the realization of the desired power to the stabilization of the receiving voltage. It should be noted that the realization of the desired power and the stabilization of the receiving voltage are similar, but slightly different. In this case, a control method unique to wireless power transmission comes into play, which I would like to introduce.

2.2.4 Short Mode

In wireless power transmission, there are various characteristics depending on the circuit topology. Sometimes it is possible to make good use of them for control and other purposes. In this section, we will discuss short mode. In the circuit shown in Figure 2.31, when $R_L = 0\ \Omega$ or $V_2 = 0$ V, short mode is selected. In other words, short mode means that the load is shorted.

From Equation (2.1), when $Z_{in2} = 0\ \Omega$, the impedance Z_2' on the secondary side viewed from the primary side becomes infinite, so ideally the current on the transmission side is 0 A and the current can be stopped. In other words, the power transmitted will also be 0 W. In reality, Z_{in2} does not become $0\ \Omega$ even when there is no imaginary component in the resonant

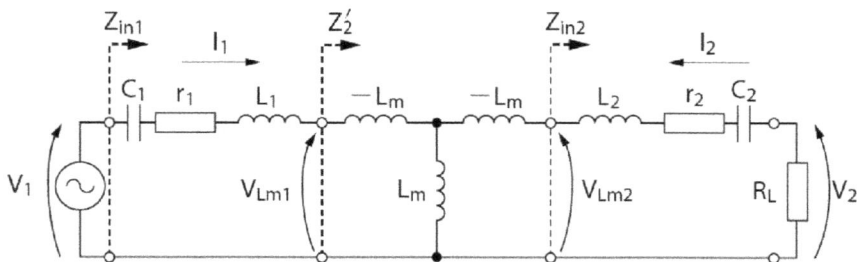

FIGURE 2.31 Short mode.

Source: [4].

state, because the internal resistance of the coil remains slightly. Therefore, Z_2' does not go to infinity, but it becomes a large value, allowing the current on the transmission side to be greatly reduced. This technique is used in an actual circuit called a HAR. This HAR is also used to control the secondary side. This will be explained in Section 2.2.5, based on the actual circuit and mode of operation.

In general circuits, power can be interrupted easily, but in wireless power transfer, energy flow control for instant termination of the primary-side power is challenging because of the separation of the primary and secondary sides. Although this control is possible with wireless communication, delays can occur and risk loss of communication with the primary side, which could damage the device in critical cases. As the system must be designed with consideration for the worst-case scenario, it is important to obtain the desired power only by controlling the secondary side. For this reason, the short mode is required.

Power is normally adjusted by primary-side voltage control. The received power can be controlled by adjusting the secondary-side power load, but the power cannot be turned off. There are, however, two modes wherein the power can be turned off in a pseudo manner or completely; these are the short mode and the open mode shown in Figure 2.32. The short mode is a realistic and effective means to achieve this requirement.

I will explain a little more with formulas. The secondary impedance at resonance Z_{in2}, the secondary impedance referred to the primary side Z'_2, and the overall circuit impedance referred to the power supply Z_{in1} are shown in Figure 2.31 and are defined by the following equations.

$$Z_{in2} = r_2 + R_L \qquad (2.31)$$

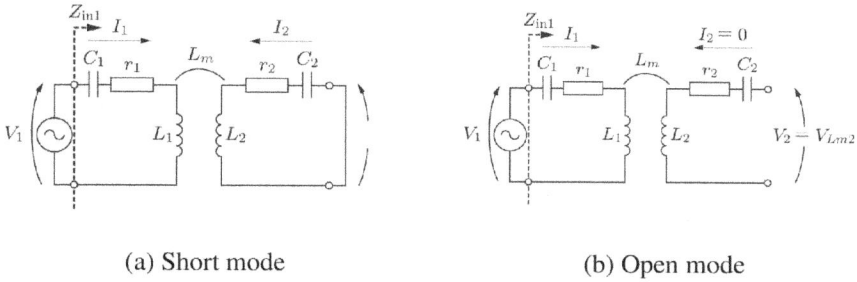

(a) Short mode   (b) Open mode

FIGURE 2.32   Short mode (left) and open mode (right).

Source: [4].

$$Z_2' = \frac{\left(\omega L_m\right)^2}{Z_{in2}} \tag{2.32}$$

$$Z_{in1} = r_1 + Z_2' \tag{2.33}$$

This implies that the primary-side current is reduced by turning off the power in a pseudo manner. Because a small current still flows on the primary side, a slight power loss occurs. On the secondary side, the current flows normally owing to the constant current characteristic. However, as the secondary-side internal resistance is low, similar to that of the primary side, the secondary-side power consumption is minimal despite the larger current. Consequently, the power can be turned off on the secondary side. This turning-off mechanism on the secondary side enables independent power control based on decisions on the secondary side, which allows many advantages.

On the other hand, in the open mode, there is no secondary-side current flow; hence, there is no power loss on the secondary side, making this mode superior to the short mode. Nevertheless, when the secondary side is open, the secondary impedance referred to the primary side Z_2' becomes very small, and a large current flows on the primary side, causing large power losses. Such a large current is problematic and not recommended for S–S. This will be further described with mathematical expressions as below. In Equation (2.31), when $R_L = \infty\ \Omega$, substitution into Equation (2.32) gives Equation (2.34). Further substituting this into Equation (2.33) gives Equation (2.35). This shows that the overall circuit impedance becomes very small, resulting in a large current.

$$Z'_2 = \frac{(\omega L_m)^2}{\infty} = 0 \qquad (2.34)$$

$$Z_{in1} = r_1 \qquad (2.35)$$

The short mode is therefore adopted and will be discussed further. Such a technology that can turn off the power whenever unnecessary, based on independent decisions on the secondary side without communication with the primary side, has many advantages and is essential for a variety of circuits such as those with constant power loads. Without this technology, the energy loses in the device will be critical and the device will be damaged.

2.2.5 HAR

So how can we achieve short mode in a real circuit [18]? Two switches are sufficient as the minimum circuit to constitute a short mode. Therefore, as shown in Figure 2.33(a), two switches are sufficient. However, if we consider bidirectional power supply, in which power is sent from the receiving side to the sending side as a further development, we may use the same circuit configuration as an inverter. In this case, as shown Figure 2.33(b), the circuit is used on the rectifier side, so it is not called an inverter but a PWM converter. Both Figures 2.33(a) and 2.33(b) have the same operation.

As shown in Figure 2.34, we simply short the two lower arms. We call it a half active rectifier (HAR) because it does not always rectify. As an operating mode, it alternates between rectifier mode and short mode. In rectifier mode, power is transmitted. And in short mode, power is cut off.

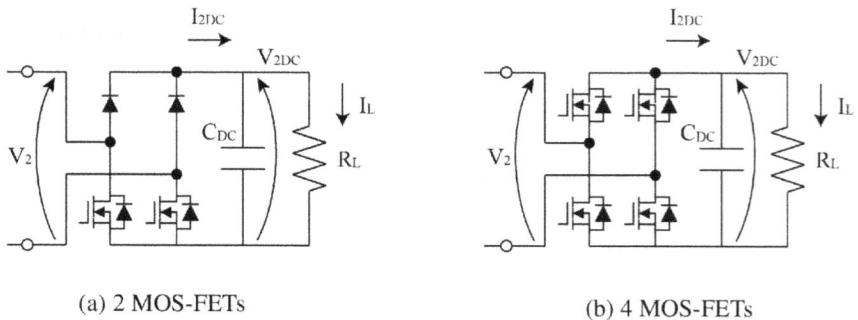

(a) 2 MOS-FETs

(b) 4 MOS-FETs

FIGURE 2.33 Circuit of HAR.

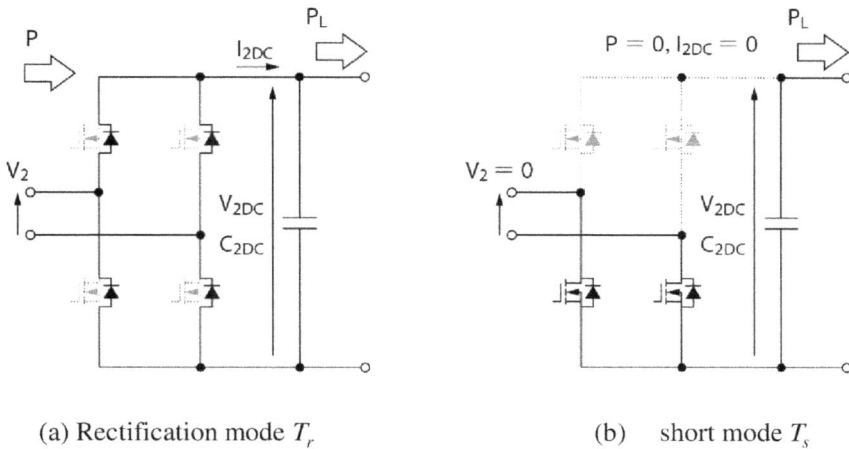

(a) Rectification mode T_r (b)   short mode T_s

FIGURE 2.34   Operation pf HAR.

Source: [4].

These two modes are used to control the power. When in short mode, power is stopped on the transmission side. Ideally, there is zero loss when in short mode, a feature unique to wireless power transmission. On the other hand, when considering the secondary side, this HAR control allows voltage control: as shown in Figure 2.35, when in rectified mode, power is delivered and the secondary-side DC link voltage rises. On the other hand, in short mode, the load uses the power, so the DC link voltage decreases. By setting threshold values for these two modes and controlling them at high speed, the secondary-side voltage can be kept constant within a certain range. In other words, the DC link voltage on the secondary side can be stabilized. Since the voltage can be controlled, either control as desired power or maximum efficiency control is possible depending on the command value. However, unless the method described in Ref. [18] is used, it is not possible to achieve both the desired power and maximum efficiency at the same time only by controlling the secondary side.

2.3 APPLICATION EXAMPLES OF WIRELESS POWER TRANSFER BY COUPLING

In this section, we will introduce examples of applications such as charging EVs while the vehicle is stationary, wireless power transfer while the vehicle is in motion, transmission through a rebar, cancer treatment, and

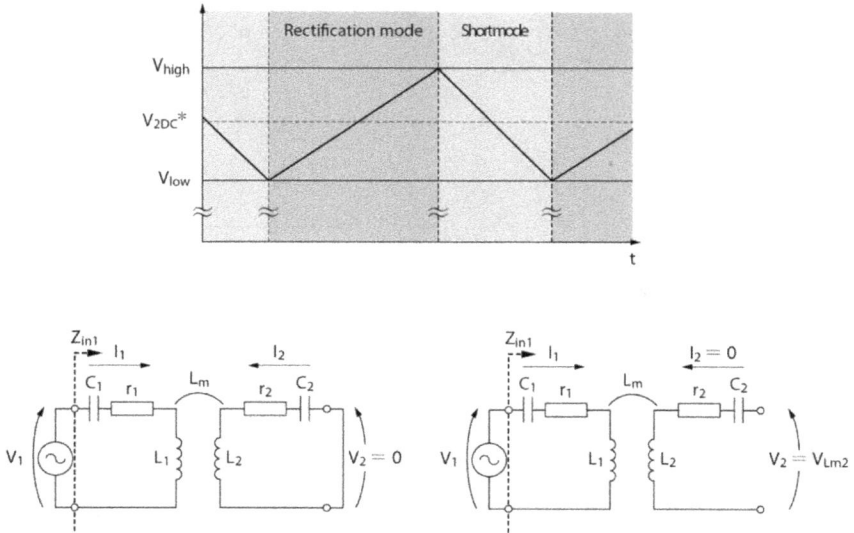

FIGURE 2.35   Secondary side voltage control by HAR.

Source: [4].

WPT in seawater. Some are still in the research stage, while others are in the process of being standardized, depending on the application.

2.3.1 Standard for EVs (SAE)

For wireless power transmission for EVs, SAE's J2954 was standardized in 2020 [19]. As for wireless power transmission using magnetic fields, IEC61980-3 and ISO19363 are moving forward in line with J2954; SAE covers both the transmission side and the receiving side (vehicle side), while IEC and ISO are divided into the transmission side and the receiving side (vehicle side), respectively. Information on J2954, which was standardized earlier, is presented here.

J2954 comprehensively describes power level categorization, three air gap categories on the transmit and receive sides, coil geometry, efficiency, circuit configuration, magnetic leakage field, and more. Among other things, J2954 focuses on and describes power level compatibility and coil shape compatibility. I will therefore introduce that. The operating frequency ranges from 79 kHz to 90 kHz. Since the efficiency of WPT is relatively robust to frequency variations, adjusting the frequency for power control does not result in a significant drop in efficiency, and thus various applications are possible, such as power control and ZVS operation by

			VA: Vehicle Assembly		
			WPT1	WPT2	WPT3
GA: Ground Assembly		WPT1	Required	Required	Required
		WPT2	Required	Required	Required
		WPT3	Required	Required	Required

FIGURE 2.36   Power level comparability.

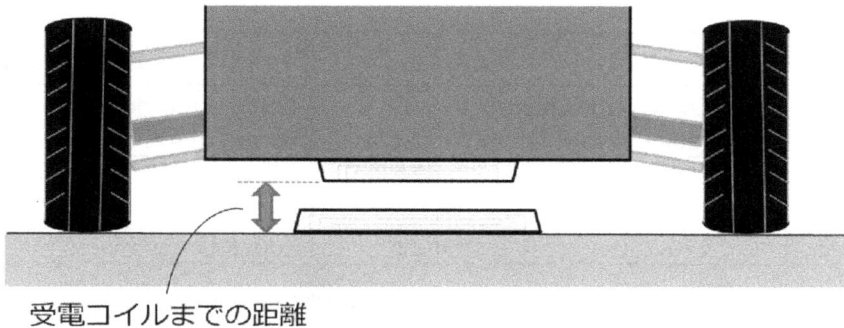

受電コイルまでの距離

FIGURE 2.37   Positioning of transmitting and receiving coils.

Source: [19].

adjusting the frequency. Therefore, various uses are possible, such as power control and ZVS operation by adjusting the frequency.

Next, power levels, there are five input power regions: WPT1 up to 3.7 kVA, WPT2 up to 7.7 kVA, and WPT3 up to 11.1 kVA. WPT4 up to 22 kVA and WPT5 over 22 kVA have been carried over to the next standard for discussion. Also, the maximum input apparent power here is specified as the power sent from the grid to the transmission equipment. Also, since this is apparent power, it is assumed to be smaller than this when converted to effective power. Although specified in kVA in the standard, it is generally rewritten and explained in kW for clarity, so care should be taken when designing strictly (Figure 2.36).

Next is the height, but it is important to note that it is not specified as the distance between the coils, but as the distance from the ground to the receiving coil pad, as shown in Figure 2.37. As shown in Figures 2.38 and 2.39, the areas to be covered by the receiving and transmitting devices are different, and although there are three categories (Z1, Z2, and Z3), the

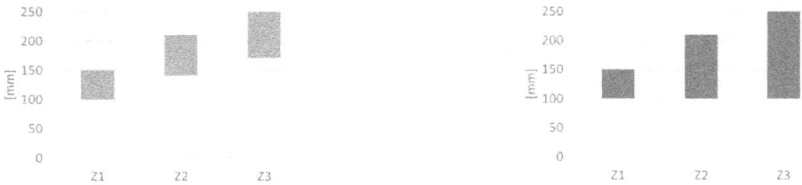

FIGURE 2.38   Graph of areas to covered by power receiving and transmission equipment.

Source: [19].

VA		GA	
Z-Class	Coil ground clearance range [mm]	Z-Class	Coil ground clearance range [mm]
Z1	100-150	Z1	100-150
Z2	140-210	Z2	100-210
Z3	170-250	Z3	100-250

FIGURE 2.39   Table of areas to be covered by power receiving and transmission equipment.

Source: [19].

area to be covered by the receiving device is somewhat limited because it varies from car to car. On the other hand, for the transmission side, it is not known which type of vehicle, i.e., which type of power receiving equipment will come, so in the case of Z3, it is required to cover all of them.

Next is the coil. A circular coil is synonymous with a spiral coil. A coil with wires wound inside, whether circular or square, is called a circular coil. A DD coil is a coil consisting of two circular coils side by side on both sides, electrically connected in series. In other words, it looks like there are two coils, but they are electrically connected, so it is not possible to turn on only one side (Figure 2.40).

2.3.2 Overview of Dynamic Wireless Power Transfer

The most famous examples of wireless power transmission while moving are transfer machines in automobile plants and semiconductor and LCD panel factories. This system, which feeds power along a lane attached to the top, was commercialized around 1993 by Daifuku Corporation based on technology from the University of Oakland. It is characterized by the

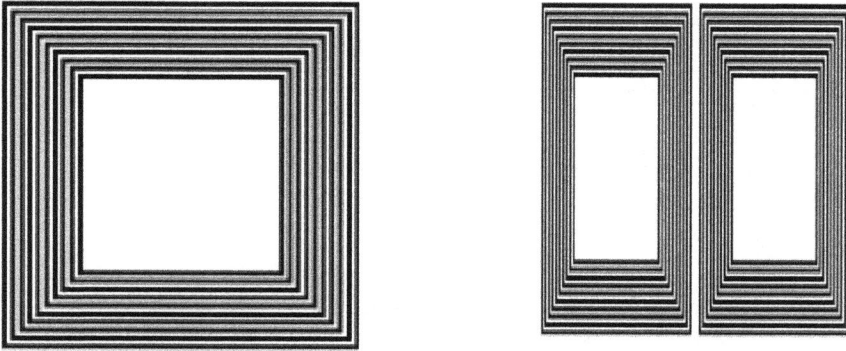

FIGURE 2.40   Circular and DD coils.

contactless connection of multiple loads to a long lane. Since the system is designed for low speed and factory use, the fluctuation parameters can be narrowed down to some extent at the time of design, such as a small air gap fluctuation range.

On the other hand, when it comes to power supply while moving these days, dynamic wireless power transfer (DWPT) for electric vehicles (EVs) is attracting attention. Although standardization discussions have begun, this technology is still in its infancy. DWPT for EVs requires a lot of variable parameters in addition to high-speed movement, and the first issue to be addressed is how much the system design can be narrowed down. The air gap varies depending on the vehicle body, and the power requirements differ between passenger cars and trucks. In addition, load fluctuations and positional deviations occur as a matter of course.

The first dividing line in designing a system is the shape of the coils. There are two types of coils: those that use long lanes (tracks) and feed power to multiple cars, and those that are divided into coils (pads type). In the case of long lanes, measures must be taken to reduce efficiency and magnetic field leakage. In addition, to feed power to multiple cars, the power required by one lane must be the same as the number of cars. For example, four cars in a lane would require 100 kW, so it is important to take countermeasures against magnetic leakage fields.

On the other hand, the coil type (pad type) improves efficiency and turns ON only when a car is present, so the body of the car has the effect of reducing the leaking magnetic field. On the other hand, for a car running through at 100 km/h, detection is made on the power transmission side on the ground side, and the system turns ON when the car arrives. Next, the

(a) lane (track) type

(b) coils (pads) type

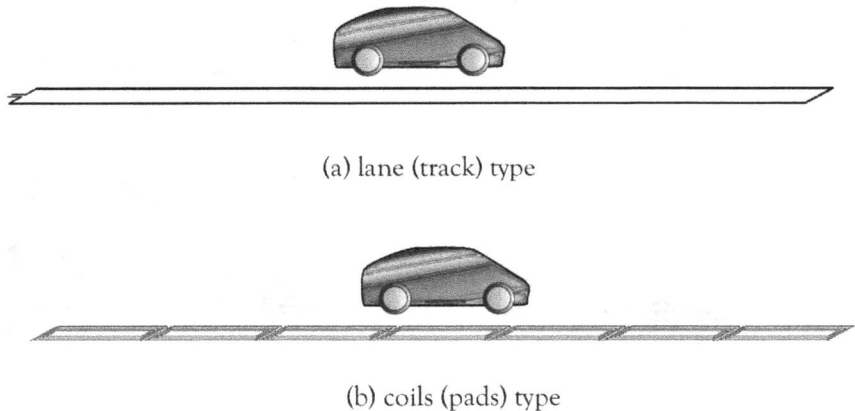

FIGURE 2.41   Coils of DWPT.

power-receiving side on the body side controls the power and efficiency. Then, when the car disappears, the power transmission side turns off. This requires advanced control and system design. Even if we assume the coil is a large one, say 2 m, it will run through in 72 ms at a speed of 100 km/h, but we are still researching and developing a system that can operate normally (Figure 2.41).

At first glance, charging while stopped and charging while driving may seem like similar technologies, but they are completely different, including the level of difficulty. The most obvious story is that in the case of charging while stopped, it is possible to control the system slowly while exchanging information using wireless communication. On the other hand, in the case of charging while driving, even if a large coil with a length of 2 m is used, only 72 ms will be spent on a single coil. Because of this situation, many new technologies are needed, such as technology that detects and sends power in a few milliseconds. Some people mistakenly believe that wireless communication is overwhelmingly fast because it travels at the speed of light, but, in reality, it is a very slow method in terms of control due to the establishment of communication, data delay, and delay in AD conversion. Therefore, wireless communication cannot be used for power control. Therefore, we need to use WPT's proprietary technology to control charging while driving without using wireless communication.

As shown in Figure 2.42, there are also various methods for the coil type: Figure 2.42(a) is the relay coil method, Figure 2.42(b) is the sensor method, Figure 2.42(c) is the sensorless active method and ground communication method, Figure 2.42(d) is the sensorless active method, and

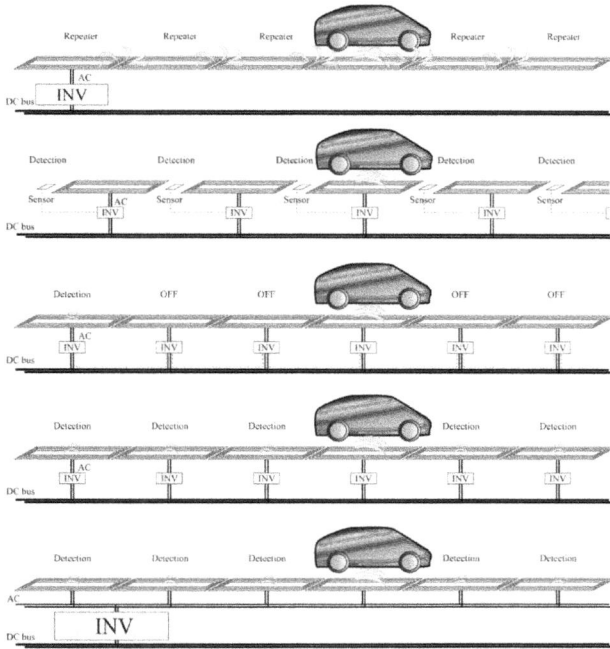

(a) Relay coil (b) Sensor method (c) Sensorless active method + ground communication (d) Sensorless active method[18] (e) Sensorless passive method

FIGURE 2.42   Various DWPT systems.

Figure 2.42(e) is the sensorless passive method. The relay coil method is easy to install, but the magnetic field leakage is large. The sensor method has a problem in case of failure. The sensorless passive method is an excellent method that automatically turns ON when a car arrives, but the loss of current for detection that is constantly generated cannot be ignored. The sensorless active method seems to be very good because all inverters perform detection, but the problem is to reduce the loss of detection current that is always generated. Adding ground communication to the sensorless active method and using the information from the previous coil seems like a smart approach, but it needs to be thoroughly verified. Thus, new technologies are needed, and in the field of wireless power transmission, research on in-transit power supply is currently a hot research topic [20].

2.3.3 DWPT System and Control

Next, we talk about control [21]. Here we use double-LCC instead of S–S in a configuration similar to Figure 2.42(e).

2.3.3.1 Outline of DWPT System and Control

DWPT systems can improve cruising range and reduce the capacity of the battery of electric vehicles. A DWPT system is required to achieve high efficiency of power transmission without complex circuits or control. In this subsection, a new DWPT system that combines sensorless energized section switching and double-LCC topology is introduced. It can transmit power safely by using double-LCC topology and efficiently by using an energized section switching system with a sensorless vehicle detection method. The effectiveness of the proposed systems was verified by experiments.

2.3.3.2 Introduction to DWPT System and Control

In DWPT systems, because of standby losses due to the internal resistance of the coils, it is desirable to turn ON only the transmission coils where the vehicle is running. Also, it is very important to guarantee safety as an infrastructure. In previous researches, it is revealed that double-LCC topology is safer than S–S (series-series) topology because of its gyrator characteristics. Here, a new DWPT system that combines sensorless energized section switching and double-LCC topology is introduced. It can transmit power safely by using double-LCC topology and efficiently by using an energized section switching system with a sensorless vehicle detection method.

2.3.3.3 Dynamic Wireless Power Transfer System

The DWPT system is classified according to the placement of the transmitter coils and the type of circuit [22,23]. The single-coil design [24], in which a very long coil is laid along the direction of the car's motion to supply power, and the segmented coil design [25,26,27], in which many coils of the same size or smaller than the car are placed on the road surface, have been studied. In order to keep the loss as low as possible and to control the magnetic field leakage to the surroundings, the segmented coil design should be adopted.

As for the circuit system, the S–S topology, S–P (series-parallel) topology, double-LCC topology, and other topologies have been proposed. The S–S topology in which compensation capacitors are connected to the transmission coils in series is shown in Figure 2.43(a), and the double-LCC system, in which LCL filters with gyratory characteristics are used, is shown in Figure 2.43(b). The inductance and capacitance on the S–S circuit are designed to satisfy the following resonance conditions with the frequency of the power supply as f_0:

(a) SS topology

(b)Double-LCC topology

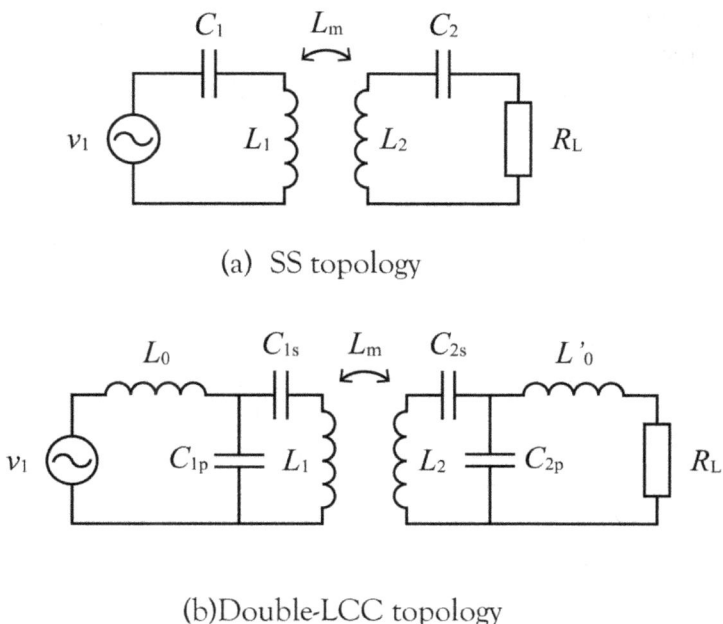

FIGURE 2.43   Equivalent circuits of WPT systems.

$$f_0 = \frac{1}{2\pi\sqrt{L_1 C_1}} = \frac{1}{2\pi\sqrt{L_2 C_2}} \quad (2.36)$$

Compared with the double-LCC topology, the S–S topology can reduce the number of elements and has the advantage of reducing the loss of energy.

The double-LCC system is designed to create LC resonance in each closed circuit. The inductance and capacitance satisfy the resonance conditions shown below.

$$f_0 = \frac{1}{2\pi\sqrt{L_0 C_{1p}}} = \frac{1}{2\pi}\sqrt{\frac{C_{1p}+C_{1s}}{L_1 C_{1p} C_{1s}}}$$

$$= \frac{1}{2\pi\sqrt{L_0' C_{2p}}} = \frac{1}{2\pi}\sqrt{\frac{C_{2p}+C_{2s}}{L_2 C_{2p} C_{2s}}} \quad (2.37)$$

When a voltage source is connected to the LCC circuit, the gyrator characteristics of the LCC circuit make it the same as when a current source is connected to the transmission coil. Therefore, double-LCC can drive multiple coils simultaneously with a single inverter. DWPT systems based on S–S and double-LCC topologies are configured as shown in Figures 2.44(a) and 2.44(b), respectively.

We compare the S–S topology and the double-LCC topology focusing on the characteristics when the coupling coefficient is zero. The primary side equivalent circuits of the S–S and double-LCC topology when the coupling coefficient is zero are shown in Figures 2.45(a) and 2.45(b). Figure 2.45(a) shows the state of the LC resonance circuit connected to the voltage source. Therefore, the impedance of the circuit as seen from the power supply is zero. Thus, in the S–S system, it is necessary to turn OFF the power supply connected to a coil that is not covered by a vehicle. This is also true for the S–P topology. On the other hand, the KVL equation of the primary side of double-LCC topology shown in Figure 2.45(b) is expressed below.

$$
\begin{bmatrix} V_1 \\ 0 \end{bmatrix} = \begin{bmatrix} j\omega L_0 - j\dfrac{1}{\omega C_{1p}} & -\left(-j\dfrac{1}{\omega C_{1p}}\right) \\ -\left(-j\dfrac{1}{\omega C_{1p}}\right) & -j\dfrac{1}{\omega C_{1p}} - j\dfrac{1}{\omega C_{1s}} + j\omega L_1 \end{bmatrix} \begin{bmatrix} I_{in} \\ I_1 \end{bmatrix}
\tag{2.38}
$$

Substitution of (2.37) into (2.38) yields (2.39).

$$
\begin{bmatrix} V_1 \\ 0 \end{bmatrix} = \begin{bmatrix} 0 & j\dfrac{1}{\omega C_{1p}} \\ j\dfrac{1}{\omega C_{1p}} & 0 \end{bmatrix} \begin{bmatrix} I_{in} \\ I_1 \end{bmatrix}
\tag{2.39}
$$

Solving this equation, input current I_{in} and primary current I_1 are derived as follows.

$$
I_{in} = 0
\tag{2.40}
$$

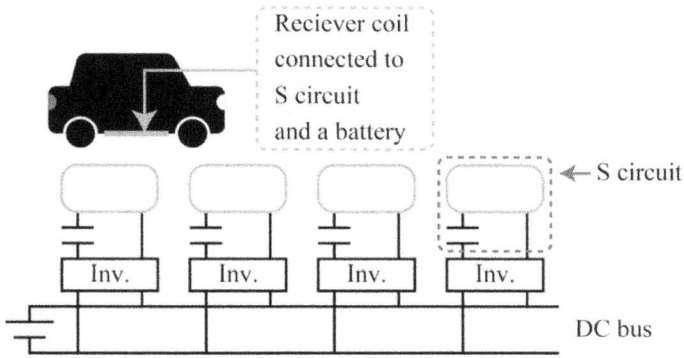

(a) SS

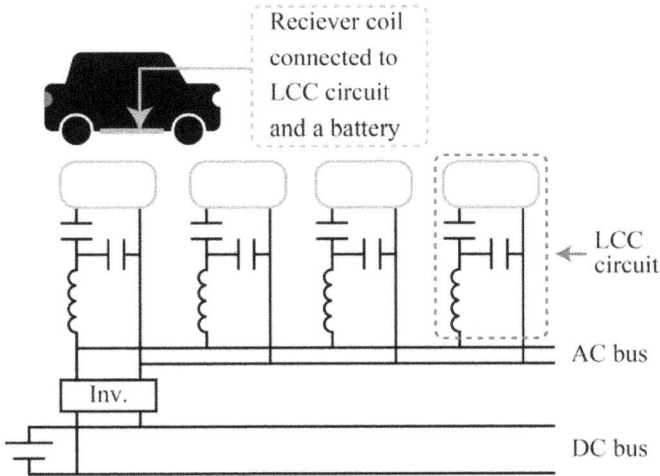

(b) Double-LCC topology

FIGURE 2.44   Illustrations of two DWPT systems.

Source: [21].

(a) SS topology

(b) Double-LCC topology

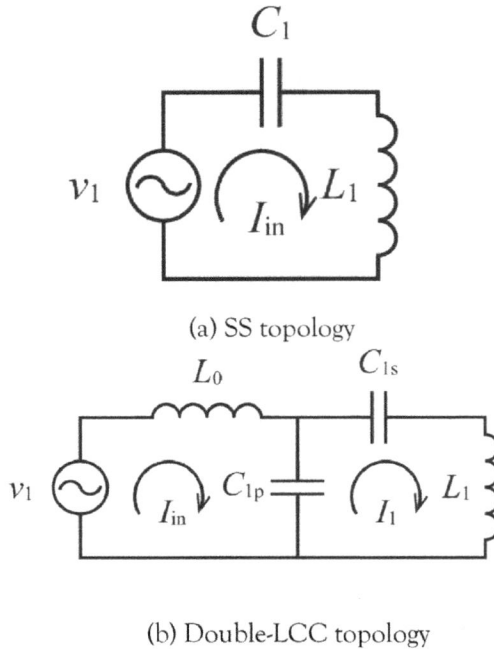

FIGURE 2.45 Equivalent circuits of primary circuits when coupling coefficient is 0.

Source: [21].

$$I_1 = -jV_1 \left(\frac{1}{\omega C_{1p}} \right)^3 \tag{2.41}$$

From Equation (2.40), no current flows from the power supply when the coupling coefficient is zero because the loop including the transmission coil L_1 is in an antiresonance state. Therefore, the output current of the inverter can be switched passively as vehicles come and go, and can be operated without external control. Since DWPT systems that serve as a social infrastructure will require extremely high safety requirements, double-LCC topology is more suitable than S–S topology. In this light, LCC-S is also suitable for DWPT systems. In LCC-S topology, it is equal to that constant voltage source connected to the load. The power source that is equivalent to the constant current source should be connected to the load because the load is a battery or an EDLC in DWPT systems. Therefore, double-LCC topology is more suitable than LCC-S topology.

2.3.3.4 Sensorless Energized Section Switching

2.3.3.4.1 System Concept In a DWPT system based on a double-LCC circuit, a group of transmission coils connected to a single inverter is collectively driven. Here, we call the section formed by this coil group a section. Because of the characteristics of the double-LCC circuit, it is possible to operate with the power supply connected to all sections always on. However, because of standby losses due to the internal resistance of the coils, it is desirable to turn ON only the sections where the vehicle is running. In order to achieve this, the ground-side system needs to detect the presence of electric vehicles and switch the sections ON and OFF at the appropriate time. In previous researches, two main topologies for detecting vehicles have been proposed. One method is to use an additional sensor such as a magnetic sensor, and the other is to use a power transmission coil as a sensor [28]. The latter is considered to be more suitable from the point of view of reducing the risk of breakdown and the cost of the system. The method using the transmission coil as a sensor is to input a signal for vehicle detection to the transmission coil and check the response of the current. The already proposed method of using the transmission coil as a sensor continues to input signals for detection even before the vehicle enters the transmission coil, and the transmission starts after the vehicle enters the transmission coil. In this system, an expensive processing unit is required for each transmission coil because the detecting procedure of EV is complex. In addition, the noise generated by the magnetic field from the transmission coils before the vehicle enters, and the lag between the vehicle's entry and the start of power transmission are also drawbacks. A system to solve these shortcomings has not yet been proposed. Therefore, a new energized section switching system using double-LCC topology is introduced. This does not apply any voltage to the coils in the section where cars do not enter and does not cause a lag between the approach and the start of power transmission.

2.3.3.4.2 Double-LCC Circuit with a Constant Voltage Load In many cases, WPT circuits are analyzed under the assumption of pure resistance as a load. However, in DWPT, as shown in Figure 2.46, a battery that acts as a constant-voltage load is connected during charging. When the input voltage is sufficiently high in Figure 2.46, the load, including the rectifier, can be approximated as a voltage source that outputs a square wave whose amplitude is equal to the DC voltage behind the rectifier and whose frequency is

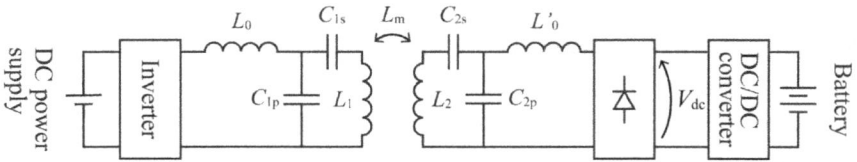

FIGURE 2.46   Double-LCC topology connected to a battery.

Source: [21].

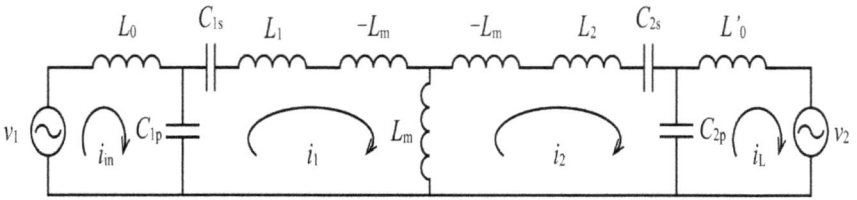

FIGURE 2.47   Analysis model of double-LCC topology connected to a battery.

Source: [21].

equal to the operating frequency of the power supply. Furthermore, since the double-LCC circuit has bandpass filter characteristics, we need to pay attention only to the basic wave. From the above, the circuit in Figure 2.46 can be simplified as shown in Figure 2.47 using the T-type equivalent circuit. The closed-circuit equation in Figure 2.47 is as follows.

$$
\begin{bmatrix} v_1 \\ 0 \\ 0 \\ v_2 \end{bmatrix} = \begin{bmatrix} Z_{11} & Z_{12} & 0 & 0 \\ Z_{21} & Z_{22} & Z_{23} & 0 \\ 0 & Z_{32} & Z_{33} & Z_{34} \\ 0 & 0 & Z_{43} & Z_{44} \end{bmatrix} \begin{bmatrix} i_{in} \\ i_1 \\ i_2 \\ i_L \end{bmatrix}
$$

(2.42)

Each term in the impedance matrix is as follows.

$$
Z_{11} = j\omega L_0 - j\frac{1}{\omega C_{1p}}
$$

(2.43)

$$
Z_{12} = Z_{21} = j\frac{1}{\omega C_{1p}}
$$

(2.44)

$$Z_{22} = -j\frac{1}{\omega C_{1p}} - j\frac{1}{\omega C_{1s}} + j\omega L_1 \tag{2.45}$$

$$Z_{23} = Z_{32} = -j\omega L_m \tag{2.46}$$

$$Z_{33} = j\omega L_2 - j\frac{1}{\omega C_{2s}} - j\frac{1}{\omega C_{2p}} \tag{2.47}$$

$$Z_{34} = Z_{43} = j\frac{1}{\omega C_{2p}} \tag{2.48}$$

$$Z_{44} = -j\frac{1}{\omega C_{2p}} + j\omega L_0' \tag{2.49}$$

Substituting the resonance condition (2.37), yields (2.50). Further deformation of the solution for i_{in} yields (2.51). Here, due to the characteristics of the double-LCC circuit, v_2 is 90 degrees ahead of v_1, so the RMS value I_1 of i_1 can be expressed as (2.52) using the RMS value V_2 of v_2.

$$\begin{bmatrix} v_1 \\ 0 \\ 0 \\ v_2 \end{bmatrix} = \begin{bmatrix} 0 & j\dfrac{1}{\omega C_{1p}} & 0 & 0 \\ j\dfrac{1}{\omega C_{1p}} & 0 & -j\omega L_m & 0 \\ 0 & -j\omega L_m & 0 & j\dfrac{1}{\omega C_{2p}} \\ 0 & 0 & j\dfrac{1}{\omega C_{2p}} & 0 \end{bmatrix} \begin{bmatrix} i_{in} \\ i_1 \\ i_2 \\ i_L \end{bmatrix} \tag{2.50}$$

$$i_{in} = -j\omega^3 C_{1p} C_{2p} L_m v_2 \tag{2.51}$$

$$I_{in} = \omega^3 C_{1p} C_{2p} L_m V_2 \tag{2.52}$$

Since V_2 is the RMS value of the basic wave of the square wave, it can be expressed as (2.53). The mutual inductance L_m can be expressed in (2.54) using the coupling coefficient k. Substituting them into (2.52), we obtain (2.55).

$$V_2 = \frac{1}{\sqrt{2}} \cdot \frac{4V_{dc}}{\pi} = \frac{2\sqrt{2}}{\pi} V_{dc} \tag{2.53}$$

$$L_m = k\sqrt{L_1 L_2} \tag{2.54}$$

$$I_{in} = \frac{2\sqrt{2}}{\pi} \omega^3 C_{1p} C_{2p} k \sqrt{L_1 L_2} V_{dc} \tag{2.55}$$

From Equation (2.55), it can be seen that the coupling coefficient can be estimated from the input current I_{in} when the constants of each passive device on the circuit and V_{dc} are constant. The basic concept of vehicle detection in this study is to detect the location of a vehicle by using the coupling coefficient estimation.

2.3.3.4.3 Energized Section Switching System As mentioned above, the double-LCC system does not allow excessive current to flow even when the coupling factor is zero, so the inverter can remain in standby with the inverter driven. By combining this characteristic with the vehicle detection system described above, the section switching system can be realized. The above operations are summarized in a flowchart as illustrated in Figure 2.48. As shown in Figure 2.48, the proposed method can be realized by extremely simple control. A current sensor connected to the rear of the section estimates the coupling coefficient. When the current flowing to this current sensor exceeds the threshold $I_{in,th1}$, which is greater than the standby current, the vehicle is judged to have entered the rear coil connected to the nth inverter. Next, when the current falls below the threshold $I_{in,th2}$, which is close to the standby current again, the vehicle is judged to have passed over the coil and entered the switching section shown by the light blue broken line in Figure 2.49. At this time, the drive of the nth inverter is stopped, and the energized section is switched by starting the drive of the $n+1$st inverter as shown in Figure 2.50. Immediately after the switchover, the system is on standby and the current begins to flow when a car comes. The standby mentioned

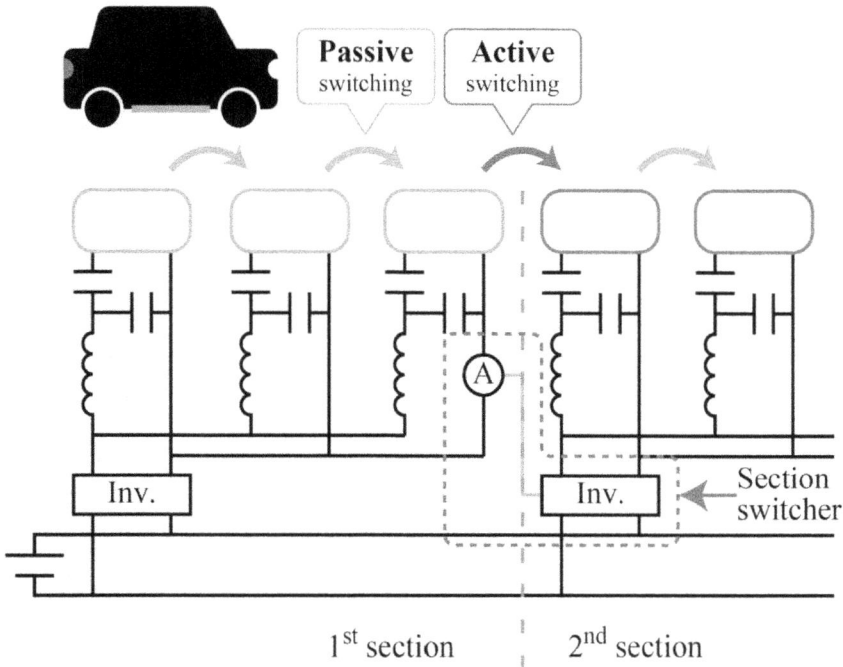

FIGURE 2.48   Illustration of energized section switching system.

Source: [21].

above can be done because it is LCC. If we do that with S–S, excessive current will flow as shown in Figure 2.51.

2.3.3.4.3 False Detection of a Vehicle   False detection of a vehicle due to errors in current sensors or signal processors may cause the inverter to start up accidentally. In the system proposed here, the double-LCC system does not cause any destruction even if a false detection occurs, and as shown in Figure 2.52, the inverter only starts up early, so that the system can return to normal power supply without any special operation. Therefore, there is no need to take special measures against false positives when using this system.

2.3.3.5 Experiment of DWPT Control

The proposed system is verified using three transmission coils (TX 1, 2, and 3), where the switching between TX 1 and 2 is active by the sensorless vehicle detection system and the switching between TX 2 and 3 is passive

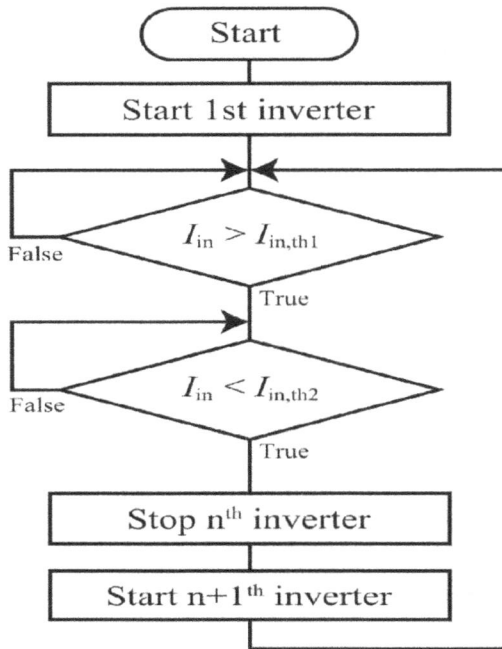

FIGURE 2.49   Flowchart of energized section switching.

Source: [21].

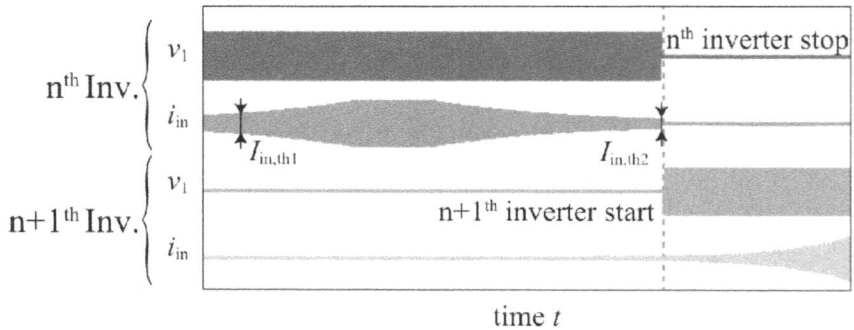

FIGURE 2.50   Concept of energized section switching on double-LCC.

Source: [21].

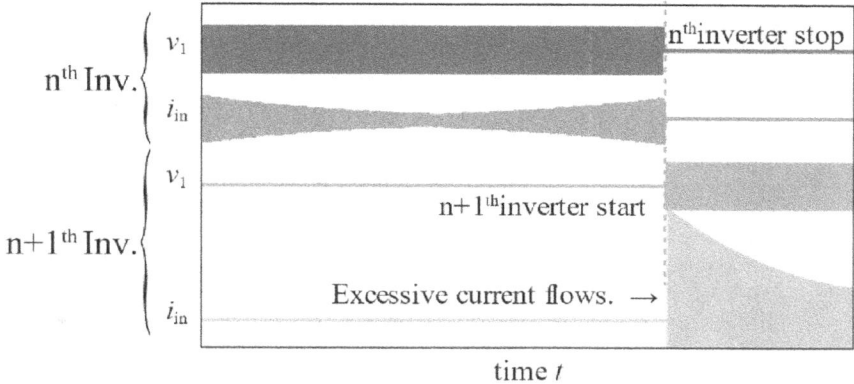

FIGURE 2.51   In the case of standby on SS topology.

Source: [21].

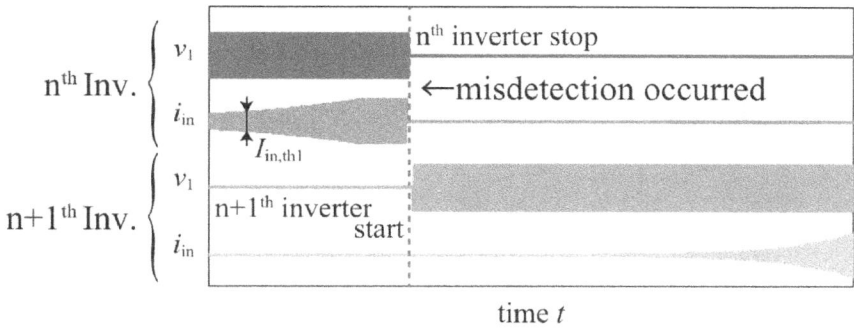

FIGURE 2.52   Concept of energized section switch when malfunction occurred, double-LCC.

Source: [21].

by the characteristics of double-LCC topology. In this experiment, TX 1 forms section 1. TX 2 and 3 form section 2. The configuration and appearance of the experimental system are shown in Figures 2.53, 2.54, and 2.55, respectively. Table 2.3 shows the parameters of the circuit used in the experiments. TX 1 is connected to inverter 1. TX 2 and 3 are connected to inverter 2. These inverters are operated by DSP on which the proposed system is implemented and output square wave voltage. The receiving coil (RX) is fixed to a platform, and the platform is driven by a servo motor at a speed of 12 km/h.

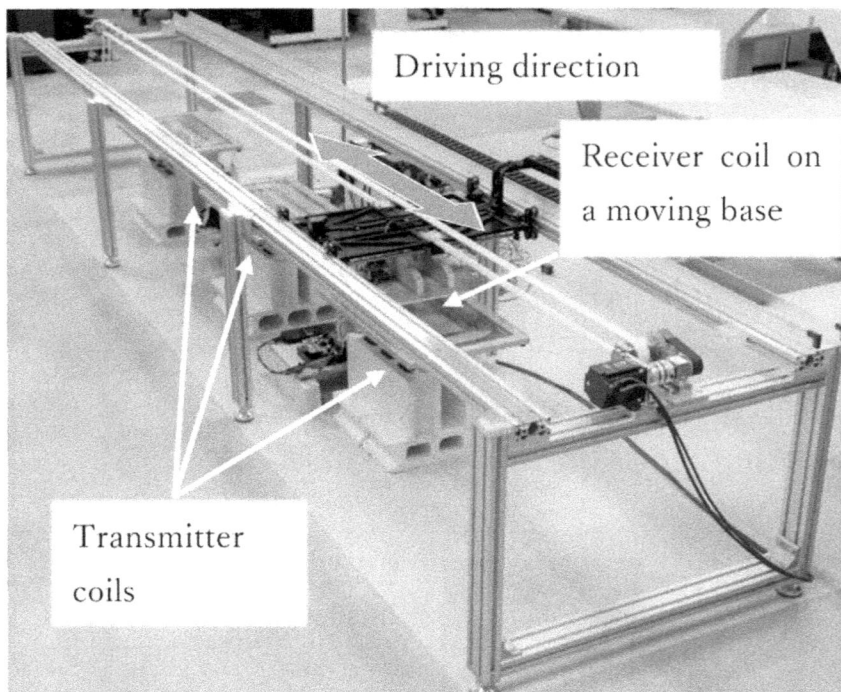

FIGURE 2.53   Experimental equipment.

Source: [21].

The output voltage and output current waveforms of inverters 1 and 2 when the system is operated with the power receiving coil running at 12 km/h using the system shown in Figure 2.54 are shown in Figure 2.56(a). An enlarged image before and after the switchover is shown in Figure 2.56(b). The output current of the inverter decreases in the area surrounded by the yellow broken line, which confirms the passive switching of the transmission coil due to the circuit characteristics of the double-LCC. After the output current of Inv. 1 decreases in the area surrounded by the red broken line, the output of Inv. 1 is stopped and the output of Inv. 2 is started. Furthermore, the output current of Inv. 2 after the switching operation is less than when the power is supplied from the transmission coil. Therefore, it is confirmed that the section switching by the current sensor is done correctly. From the above, it can be seen that the proposed method can switch the TX 1, 2, and 3 as the receiver coil moves.

TABLE 2.3 Experimental Parameters

		TX 1	TX 2	TX 3	RX
Input voltage	V_{in}	5 V_{rms} (square wave)	5 V_{rms} (square wave)	5 V_{rms} (square wave)	—
Load voltage	V_{dc}	—	—	—	2 V
Operation frequency	f	85 kHz	85 kHz	85 kHz	—
Transmitter size		500×250 mm	500×250 mm	500×250 mm	—
Receiver size		—	—	—	250×250 mm
Compensated inductance	L_0, L_0'	20.671 μH	20.811 μH	21.094 μH	12.003 μH
Compensated coil resistance	r_0, r_0'	30.669 mΩ	29.517 mΩ	41.698 m'	19.127 mΩ
Compensated capacitor	C_{1p}, C_{2p}	164.97 nF	164.97 nF	164.97 nF	289.55 nF
Resonant capacitor	C_{1s}, C_{2s}	18.823 nF	18.823 nF	18.823 nF	35.367 nF
Transmitter/receiver inductance	L_1, L_2	206.88 μH	206.68 μH	206.43 μH	111.44 μH
Transmitter/receiver resistance	r_1, r_2	148.78 mΩ	125.40 m'	129.93 mΩ	79.459 mΩ

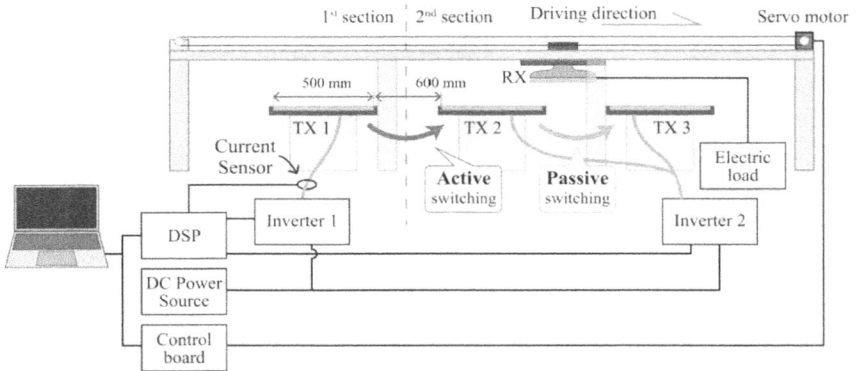

FIGURE 2.54   Configuration of experiment.

Source: [21].

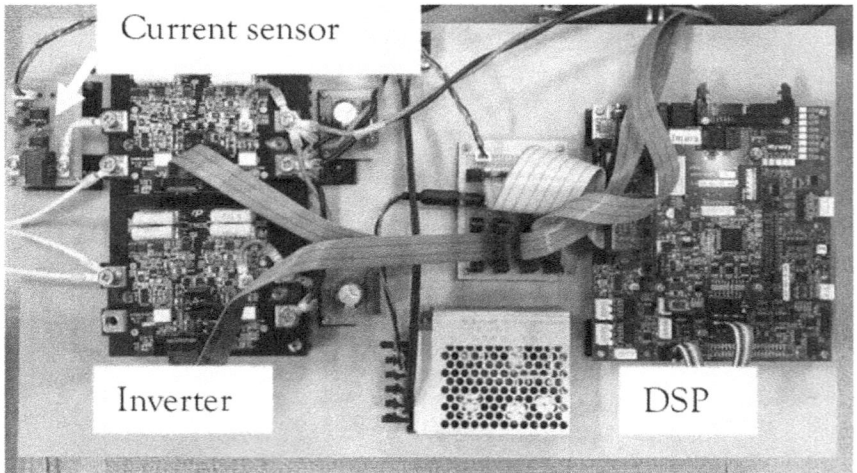

FIGURE 2.55   Control board of experimental equipment.

Source: [21].

2.3.3.6 Conclusion of DWPT Control

In this subsection, a new energized section switching system that does not apply any voltage to the coils in the section is introduced where cars do not enter and do not cause any lag between the approach and the start of transmission. In the proposed system, sensorless energized section switching and double-LCC topology are combined. It can transmit power safely by using double-LCC topology and efficiently using an energized section switching

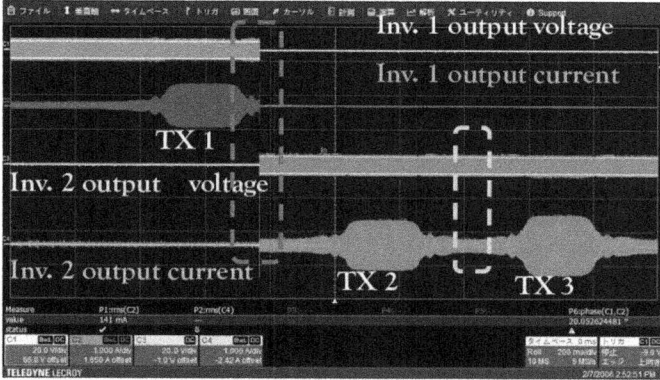

(a) Overall view

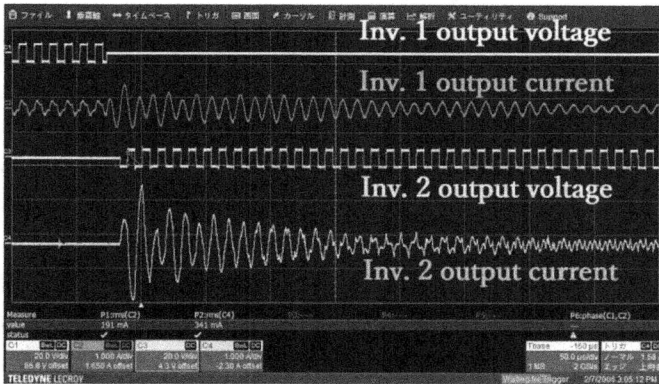

(b) Enlarged view

FIGURE 2.56   Outputs of inverters when switching system is running.

Source: [21].

system with a sensorless vehicle detection method. The proposed method requires only primary current sensing and can be implemented by a simple processing unit. Also, since additional sensors such as magnetic sensors are not used, the transmission system can be manufactured with relatively simple circuits, which reduces the risk of breakdown. Therefore, it can be said that the proposed method is suitable for a transmission system that is buried in the ground for a long time. Experiments have confirmed that the power transmission coils switch following the movement of the receiving coils, demonstrating that the system can be operated during driving.

2.3.4 Coil-embedded Road for DWPT

DWPT [29] has another very important research theme in addition to control technology that can cope with high-speed running. This is the study of coil installation technology. Many people think that coils are simply buried in the road, but in reality, it is not that simple. Here, we introduce our research on coil burial in asphalt.

2.3.4.1 Outline of Coil-embedded Road for DWPT

The development of technology for embedding coils in roads is important for dynamic wireless power transfer (DWPT). When embedding coils in a road, it is necessary to ensure both the electrical and mechanical properties required for DWPT. In this study, coils are produced using PP and PE, which are synthetic resins. In addition, the coil is mainly an open-type coil without the ferrite and capacitor. The electrical characteristics of the coil were evaluated before and after burying it. Further, we evaluated the mechanical strength of the road before and after the coils were buried using the falling weight deflectometer test. Four construction methods were investigated. The deformation of the inside of the coil was measured using strain gauges, and during paving, the effect of the heat of the asphalt in the space inside the coil was measured using thermocouples. A comprehensive evaluation of the electrical and mechanical properties of the synthetic resin coils was conducted when they were paved. The results show that the reflection crack suppression sheet (RC mesh) method is currently the best construction method. Moreover, it was found that injecting cement grout to protect the coil was the best way to reduce the residual strain.

2.3.4.2 Introduction to Coil-embedded Road for DWPT

Global warming is an unavoidable issue that must be considered by the transportation sector. DWPT [30–40] has attracted significant attention as a sustainable system to reduce CO_2 emissions in the transportation sector. There are two types of DWPTs: magnetic field coupling and electric field coupling. The magnetic field method is used because it is more resistant to rain. The cost of the coils accounts for one-fourth of the total cost of equipment and construction for the installation of the DWPT system, and if the cost can be reduced, the economic viability of DWPT will increase. In this study, ferrite-less and capacitor-less coils [41–43], which are low-cost coils, are embedded in roads to achieve the requisite electrical and mechanical properties. Four construction methods have been tested. The electrical characteristics, efficiency, and power are verified. As for the mechanical properties, we verified the strength of the coil itself using strain gauges. Furthermore, the effect of heat from the asphalt mixture generated during pavement construction was studied. The structural strength of the road pavement was evaluated by a deflection measurement test using a falling weight deflectometer (FWD). Then, after verifying the bearing capacity of cracks on the road, the service life of the road was calculated. Finally, a comprehensive evaluation of the electrical and mechanical properties was conducted.

2.3.4.3 DWPT System Configuration
The following subsections describe the overall project and the DWPT system.

2.3.4.4 DWPT Road and Measurement Environment
The experiment was conducted by burying the coil in a section of a 110 m DWPT road at the Noda campus of Tokyo University of Science. An aerial view of the test site is shown in Figure 2.57. In this study, only coil-to-coil evaluations were performed to measure the characteristics accurately, although the environment here allows experiments at speeds of up to 70 km/h. The measurement environments of the transmitting and receiving coils before and after the burial are shown in Figure 2.58.

2.3.4.5 Four Construction Methods: Case 1 to Case 4
The four construction methods shown in Figure 2.59 are described as follows: Case 1: covering the coils with an RC mesh sheet; Case 2: placing unreinforced cement concrete retaining plates under the coils and covering the coils with a reflection crack suppression (RC mesh) sheet; Case 3: wrapping the coil horizontally and vertically with stretch film and winding it vertically and horizontally three times; and Case 4: injecting cement grout around the coil. The transmitting coil sizes and types are listed in Table 2.4. A list of construction methods is presented in Table 2.5.

FIGURE 2.57 Aerial view of running feeder section. This figure was created by processing an aerial photograph taken by the Geospatial information Authority of Japan (taken in 2013).

Source: [29].

FIGURE 2.58 Transmitting and receiving coils' measurement environments before and after burial.

Source: [29].

The power-receiving coil is shown in Figure 2.60. The parameters used are listed in Table 2.6. The Q-value of the power-receiving coil is over 200, although it has not been optimized.

2.3.4.6 Coil-embedded Road Construction

The flow of the coil burial work is shown in Figure 2.61. Cutting and pre-treatment of the pavement are necessary for embedding a coil in an

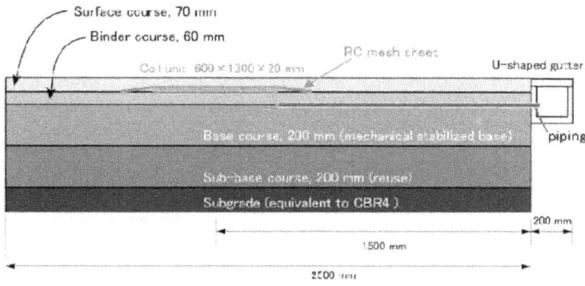

(a) RC mesh sheet

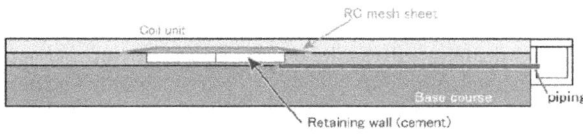

(b) Retaining wall (cement) and RC mesh sheet

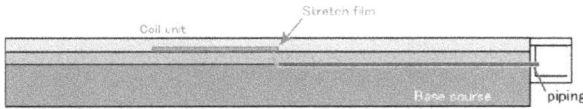

(c) Stretch film

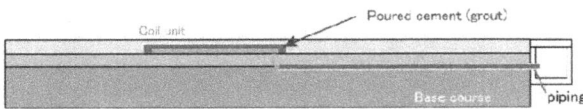

(d) Cement (grout)

FIGURE 2.59   Four methods of coil burial.

Source: [29].

TABLE 2.4   Power Transmission Coil Specifications

Coil size (case size)	600 × 1300 mm (694 × 1380 × 20 mm)
Turns	36
Interlayer gap	1 mm
Line pitch	7 mm
Conductor diameter	3 mm
Weight	PP: 22.5 kg, PE: 23.6 kg

TABLE 2.5   List of Construction Methods

Number of cases	Construction method	Material
Case 1	RC mesh	PP (top and bottom layer)
Case 2	Retaining wall and RC mesh	PP (top and bottom layer)
Case 3	Stretch film	PP (top and bottom layer)
Case 4	Cement (grout)	PE (top and bottom layer)
Common feature	–	PE (middle layer)

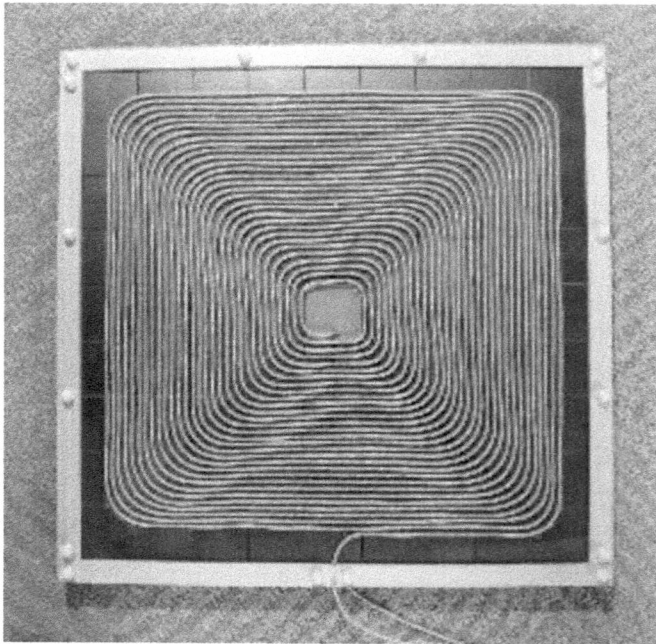

FIGURE 2.60   Power-receiving coil.

Source: [29].

TABLE 2.6   Power-receiving Coil Specifications

Coil size	400×400 mm
Turns	30
Line pitch	6 mm
Conductor diameter	3 mm
Total length of line	$\fallingdotseq 30$ m
Weight	8.57 kg

(a) Cutting   (b) Removal of asphalt mixture

(c) Base course (MSCS)   (d) Piping setup

(e) Binder course   (f) Finished binder course

(g) Coil installation 1   (h) Coil installation 2

FIGURE 2.61   Flow of coil burial work.

Source: [29].

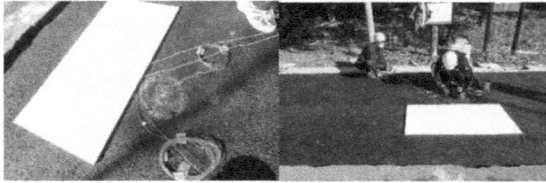

(i) Before wiring   (j) Wiring work

(k) Complete coil installation and wiring using the four methods

(l) Paving of surface course   (m) Finished pavement

FIGURE 2.61 (Continued)

(a) Laying of RC mesh   (m) Finished pavement

FIGURE 2.62 Case 1: RC mesh.

Source: [29].

(a) Soil retaining board

(b) Soil-retaining board after burial

(c) Coil installation

(d) Completed RC mesh

FIGURE 2.63   Case 2: soil retaining board and RC.

Source: [29].

(a) Film winding

(b) Completion

FIGURE 2.64   Case 3: Stretch film.

Source: [29].

(a) Cement grout into formwork

(b) Surface finish

(c) Removal of formwork.

(d) Finished

FIGURE 2.65   Case 4: Grout material, cement for pouring.

Source: [29].

existing road. After laying the base course (mechanically stabilized crushed stone) and installing the piping, binder course work was performed. The coils were then installed and wired, and four construction methods were performed. Finally, the pavement was completed by paving the surface of the asphalt mixture. The details of the four construction methods are shown in Figures 2.62–2.65.

After the coils were embedded, they were painted, as shown in Figure 2.66, to show where they were embedded.

2.3.4.7 Electrical Characteristics of Coils

In this section, we discuss the electrical characteristics of the system. The main coil used in this study is an open-type coil, which is called the ferrite-less and capacitor-less coil. The open-type and short-type coils are shown in Figure 2.67. The open-type coil uses self-resonance owing to stray capacitance, and thus does not require an external resonant capacitor. Moreover, it does not require ferrite, because it is sufficiently efficient. Therefore,

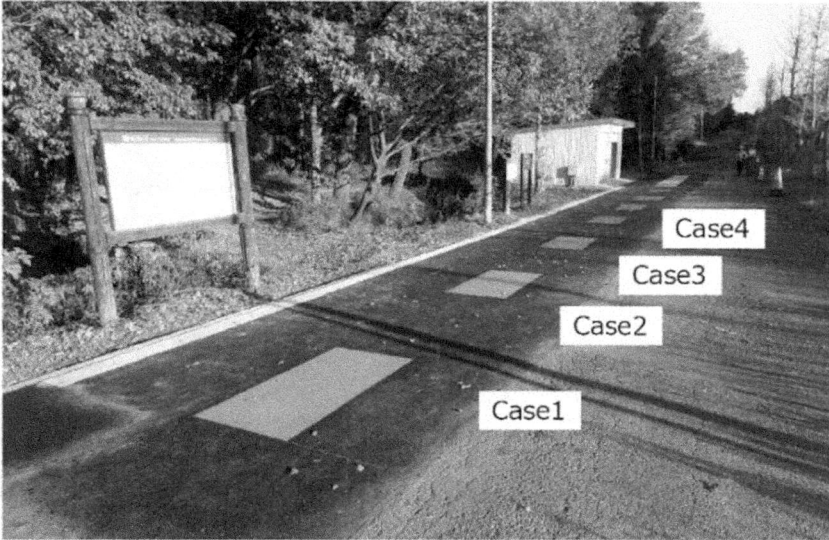

FIGURE 2.66   Post-construction painting of the four cases.

Source: [29].

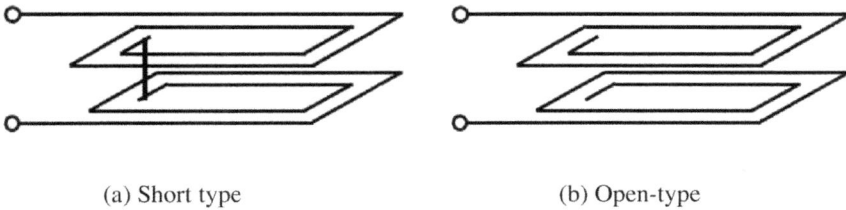

(a) Short type   (b) Open-type

FIGURE 2.67   Schematic of open and short coils.

Source: [29].

only a Litz wire is needed to form the coil, which enables cost reduction compared to conventional short-type coils.

2.3.4.8 Case Selection for DWPT Buried Open Coil

This open-type coil is made of Litz wire and case only, and has a three-layer structure with grooves for winding the coil on the upper and lower layers. There is an intermediate layer between the upper and lower layers to secure the separation distance between the upper and lower layers. The middle layer was composed of PE (Figure 2.68).

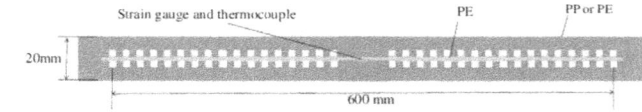

(a) Sectional view

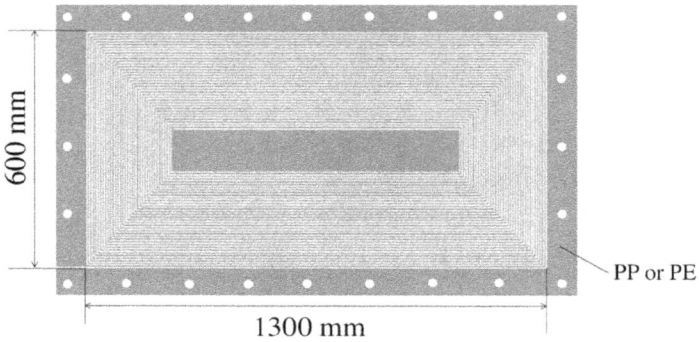

(b) Top view

FIGURE 2.68   Coil structure.

Source: [29].

TABLE 2.7   Candidate Coil Case Materials

	PE	PP	ABS	XPS	PC
tanδ (1 kHz)	Excellent 0.0005	Good 0.0005–0.0018	Fair 0.004–0.007	Excellent 0.0001–0.0003	Poor 0.021
Bending strength (kg/cm^2)	Poor 7	Good 422–562	Excellent 696–914	Poor 20	Excellent 949–960
Compression strength (kg/cm^2)	Poor 199–253	Good 387–562	Excellent 738–879	Very poor 16	Excellent 780–879
Deflection temperature under load (℃)	Poor 43–54	Poor 49–60	Good 93–107	Good 80	Excellent 132–137
Insulation strength (kV/mm)	Good 450–500	Excellent 500–660	Fair 350–500	Excellent 500–700	Fair 400
Material cost (1000 × 2000 × 12 mm)	Poor 54,890 yen	Good 11,495 yen	Fair 43,494 yen	Excellent 1362 yen (910 × 1820 × 20 mm)	Fair 38,000 yen

Open-type coils use floating capacitance, so the electric field component near the Litz wire is larger than that near the short type. Therefore, it is susceptible to dielectric loss due to the tan δ of the coil. A material with low tan δ will have high efficiency and high power. It is also necessary to consider mechanical strength and cost. The characteristics of typical resins are shown in Table 2.7.

ABS and polycarbonate (PC) are not candidates because they have large tan δ values and are expected to adversely affect efficiency. Extruded polystyrene foam (XPS) is not a candidate because its mechanical strength is very poor.

From the above, we adopt PP and PE because their mechanical strength is not poor, and their efficiency and power are expected to be high. Because the tan δ values of the PP and PE materials were almost the same, PE was also used in this experiment. PP is used in construction methods 1–3, and PE is used in construction method 4.

(a) One-sided 1 layer (PP)

(b) Coil structure (PE)

(c) Finished form (PP)

(d) Completed form (PE)

FIGURE 2.69   Photograph of open-type coil.

Source: [29].

A photograph of the coil used is shown in Figure 2.69. As mentioned earlier, the structure is the same for all the coils and consists of three layers: the top, middle, and bottom layers.

2.3.4.9 Coil Electrical Characteristics before and after Burial

The locations of the sending and receiving coils and the relationship between the surface course and binder course are shown in Figure 2.70. The distance between the coils was 140 mm, and the gap between the ground and the receiving coil was 90 mm.

The parameters of the coils measured in the room before and after embedding the coils in the road are listed in Table 2.8. In general, the resistance increased and the Q-value decreased after the coil was

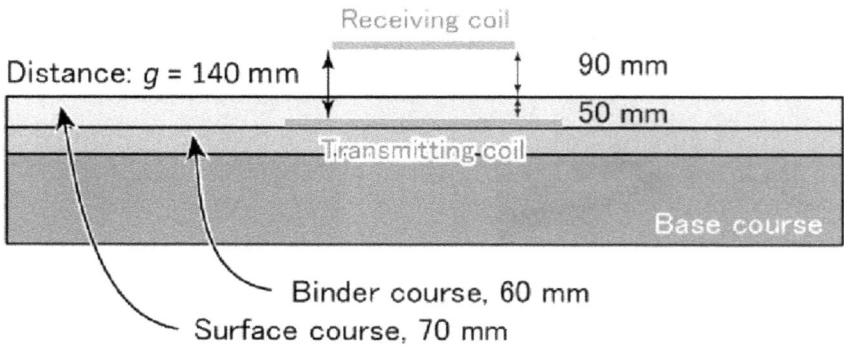

Distance: g = 140 mm   90 mm

Receiving coil

50 mm

Transmitting coil

Base course

Binder course, 60 mm
Surface course, 70 mm

FIGURE 2.70   Distance between transmitting and receiving coils and road structure.

Source: [29].

TABLE 2.8   Coil Characteristics before and after Burial

Construction method	Timing	f_0 (kHz)	R (Ω)	Q(= ωL/R)	L (mH)	C (nF)
Case 1	Before	76.2	2.71	391	2.21	19.7
	After	64.6	5.27	191	2.49	2.44
Case 2	Before	69.8	2.43	411	2.27	2.29
	After	69.8	30.9	25.7	1.81	2.87
Case 3	Before	84.8	2.61	438	2.15	1.64
	After	60.8	6.97	132	2.41	2.84
Case 4	Before	78.6	2.33	432	2.04	2.23
	After	60.4	14.6	59.1	2.27	3.05

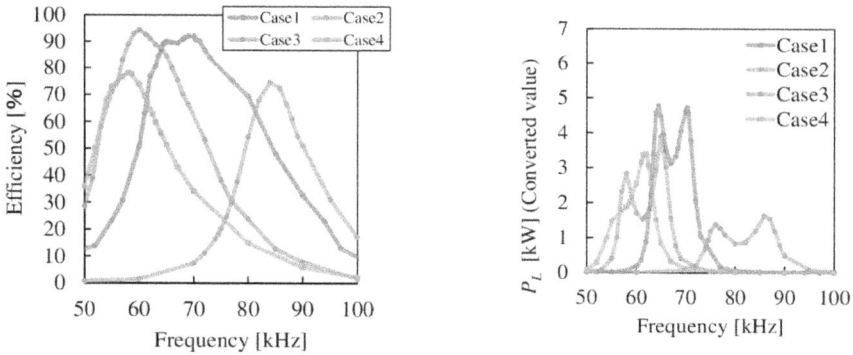

FIGURE 2.71   Efficiency and power against frequency after the coils are buried.

Source: [30].

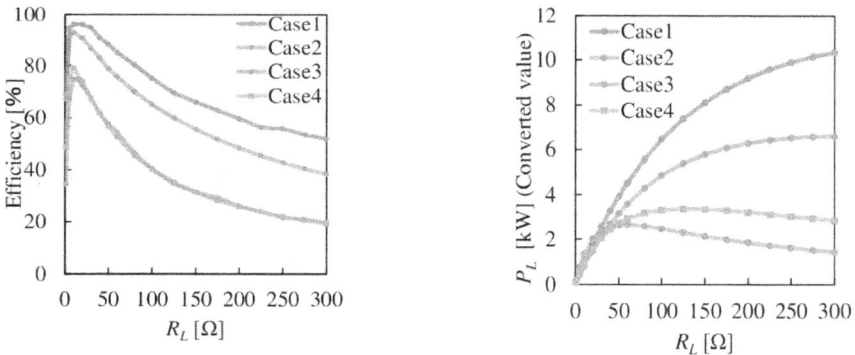

FIGURE 2.72   Efficiency and power against load after the coils are buried.

Source: [30].

embedded, and the coil characteristics deteriorated. In particular, the Q-values of cases 2 and 4 using cement deteriorated significantly to less than 100. By contrast, the Q-values of cases 1 and 3 were above 100, although they deteriorated.

Figure 2.71 shows the efficiency and power against frequency, and Figure 2.73 shows the efficiency and power against the load. Tables 2.9 and 2.10 show the efficiency and power before and after the coils were buried in the road, and Table 2.11 shows the power before and after the coils were buried in the road at 85% efficiency. The resonant frequency deviates from 85 kHz because the stray capacitance deviates from the design before the coil is embedded. As for the effect of coil embedment under the road, it can be said that the frequency changes significantly when dealing with

(a) Overview   (b) Strain gauge (left) and thermocouple (right)

FIGURE 2.73   Position of stain gauge and thermocouple in the coil.

Source: [30].

TABLE 2.9   Efficiency and Power before Coils are Buried

	Case 1	Case 2	Case 3	Case 4
Frequency (kHz)	68.8	67.5	64.0	72.3
R_{Lopt}	18	18	16	21
η_{max} (%)	97.1	97.1	96.6	97.1
$P_{L,\eta max}$ (kW) (converted value)	2.3	2.3	2.3	2.5

TABLE 2.10   Efficiency and Power after Burial

	Case 1	Case 2	Case 3	Case 4
Frequency (kHz)	68.4	85.8	61.3	58.6
R_{Lopt}	10	15	10	10
η_{max} (%)	96.0	73.6	92.8	78.9
$P_{L,\eta max}$ (kW) (converted value)	2.7	1.4	1.4	1.4

TABLE 2.11   Power before and after Burial (Converted Value)

Construction method	Timing	>85%
Case 1	Before	14.1 kW
	After	4.8 kW
Case 2	Before	14.5 kW
	After	-
Case 3	Before	15.5 kW
	After	3.4 kW
Case 4	Before	15.7 kW
	After	-

cementitious materials in Cases 2 and 4. In addition, the burial of the coil under the road reduces efficiency and power. The maximum efficiency was over 98% in Case 1, but the power was less than 3 kW, and the efficiency was over 91% in Case 3. Cases 2 and 4 are construction methods using cement-based materials, and their efficiency is less than 80%. Considering the case where the efficiency is more than 85%, the power was more than 14 kW before the coils were buried, but after the coils were buried, the power was 4.8 kW for Case 1. The power values were converted to 600 V input.

2.3.4.10 Mechanical Properties of Coils

The mechanical properties of the coils were evaluated to check the effect of distortion caused by rolling during road paving and the effect of heat from asphalt exceeding 100°C during paving. Figure 2.73 shows a photograph of the strain gauges and thermocouples installed in the center of the coil in the first layer. The position of the thermocouple on the roadside is shown in Figure 2.74.

The internal temperature of the pavement while adding asphalt is shown in Figure 2.75. The surface and base course were above 100°C, while the asphalt surface was above 145°C. The base course and subgrade were not significantly affected by the heat. Figure 2.76 shows the temperature of the coil. The temperature of the coil is affected by the load deflection temperature and needs to be improved.

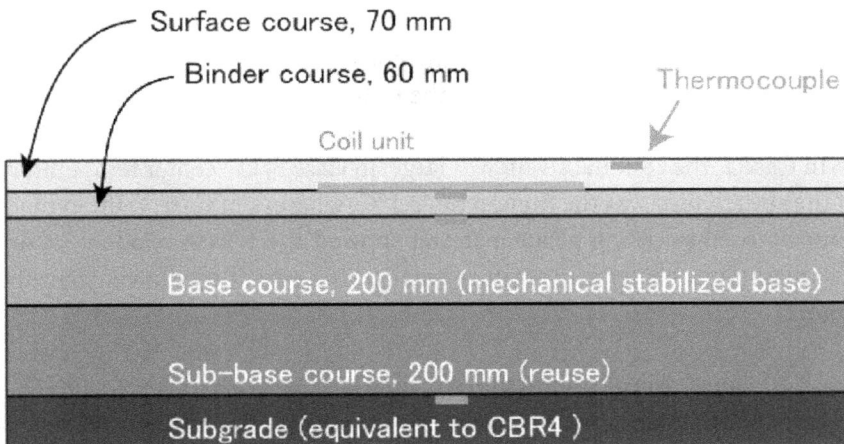

FIGURE 2.74   Position of thermocouple in the road.

Source: [29].

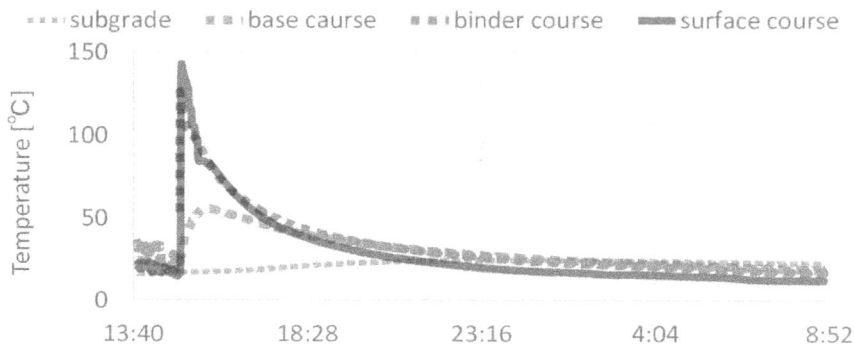

FIGURE 2.75   Internal temperature of the pavement.

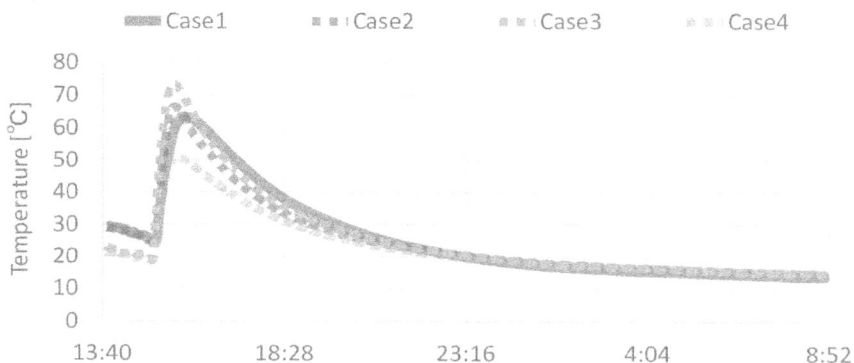

FIGURE 2.76   Temperature at the center of the coil unit.

Figure 2.77 shows the strain of the coils. During the paving work, the distortion is as expected because of the rolling pressure from the roller, but after the laying work, residual strains were observed.

In Case 1, the residual strain was large; in Case 3, the center temperature of the stretch film was the highest at 73.1°C, whereas in Case 4, the cement pouring method using grout material showed the lowest residual strain. The maximum center temperature was 50.8°C, which was lower than the others.

2.3.4.11 Mechanical Characteristics of Road Pavement

Regarding the mechanical properties of the road, the amount of deflection from the FWD test and the number of years the road can be used, calculated from the FWD test, will be verified.

2.3.4.12 Impact on Roads

The FWD test was used to evaluate the bearing capacity with respect to the cracks. The FWD test scene is shown in Figure 2.78. The results of the deflection measurement of the road pavement by FWD are shown in

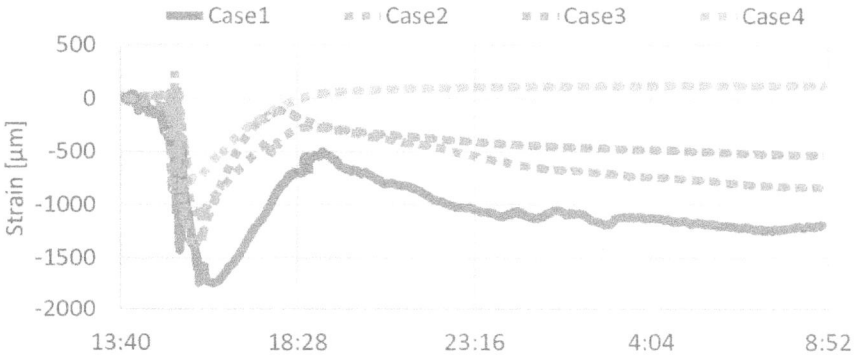

FIGURE 2.77   Strain at the center of the coil unit.

Source: [29].

(a) Positioning

(b) FWD test vehicle and sensors

(c) Deflection sensor

(d) Falling weight

FIGURE 2.78   Deflection measurement test by falling weight deflect meter (FWD) test.

Source: [29].

Figure 2.79. The deflection in Cases 1 and 3 is 1.1 mm. The deflection of a typical pavement without coils is 0.8 mm.

The deflection of Case 2 with unreinforced concrete soil-retaining plates was 1.85 mm, and that in Case 4 where coils were covered with cement grout was 1.36 mm, which is larger than that of the general section without coils. It is presumed that the lack of adhesion between the asphalt mixture and cementitious material caused a larger strain inside the coil.

2.3.4.13 Allowable Driving Years

The allowable driving years calculated from the FWD test are listed in Table 2.12. The N5 traffic volume is assumed to be 250–1000 heavy vehicle

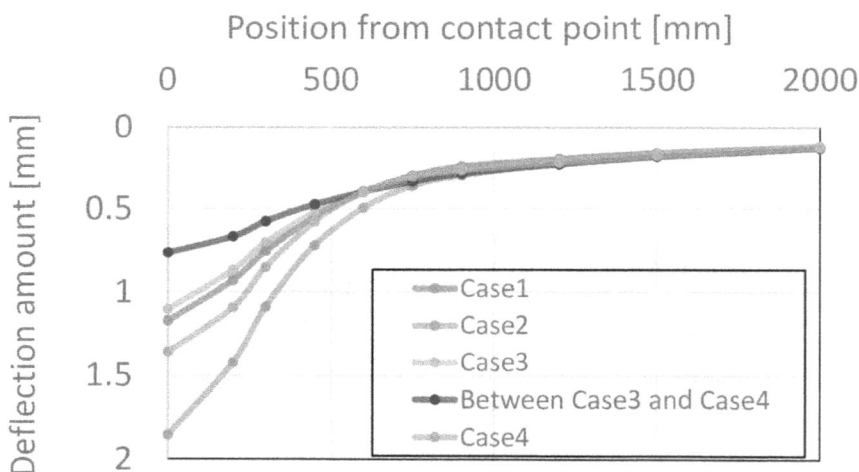

FIGURE 2.79   Deflection of road pavement.

Source: [29].

TABLE 2.12   Number of Years the Road Can Be Traveled

	Case 1	Case 2	Case 3	Case 4
Limitation on the number of loaded wheels (17 kN)	3,423,500	698,600	3,925,500	1,668,100
Number of trip limits (17 kN)	140,026,472	28,573,826	160,559,052	68,227,883
Limitation on the number of loaded wheels (49 kN)	77,800	36,900	71,200	46,100
Number of trip limits (49 kN)	3,182,141	1,509,267	2,912,191	1,885,562
Road life (years) (N5)	34.2	16.2	31.3	20.3
Road life (years) (N6)	6.8	3.2	6.3	4.1

traffic per day, and the N6 traffic volume is assumed to be 1000–3000 heavy vehicle traffic per day. A total of 255 vehicles for N5 and 1275 vehicles for N6 were used as representative values in the calculation.

In Cases 1 and 3, the number of fatigue fracture wheels is more than 70,000 cycles with a design wheel load of 49 kN, and more than 3.4 million cycles with a design wheel load of 17 kN for small roads. The general section without coils can tolerate more than 365,000 cycles at a design wheel load of 49 kN, and the value with coils is approximately five times that without coils.

If the operation is between wheels with a few wheels passing positions, the probability of a wheel passing over the coil is 2.4% [44]. Considering this, the allowable driving years is expected to be 16–34 years for the N5 traffic equivalent road. It is also judged that the system can be in service for more than six years for N6 traffic volume equivalent road.

From Table 2.12, the overall evaluation of the electrical and mechanical properties of the coil and the mechanical properties of the road show that Case 1, Case 3, Case 4, and Case 2 are superior, in that order. However, even in Case 1, there is still room for improvement because the problem of residual strain has not been solved and the coil cannot be used for 10 years at the equivalent of N6 traffic volume.

Table 2.13 presents the overall evaluation of the five stages. The overall evaluation is unfavorable for Cases 2 and 4 because of their low efficiency. However, Case 4 was promising because of its ability to reduce residual strain. Case 3 is the cheapest in terms of cost but has no outstanding features. Although Case 1 has some issues with residual strain, it has the highest overall rating because of its high efficiency, large albeit insufficient power, and high road pavement durability.

TABLE 2.13   Construction Method Five-step Comprehensive Evaluation

Category	Evaluation	Case 1: RC mesh	Case 2: Retaining wall and RC mesh	Case 3: Stretch film	Case 4: Poured cement (grout)
Electrical characteristics, coil	Efficiency	5	2	4	2
	Power	3	1	1	1
Mechanical properties, coil	Temperature	4	4	3	5
	Residual strain	2	3	3	5
Mechanical properties, road	Durability (FWD)	3	1	3	2
Other	Workability	5	3	3	4
Total	Overall evaluation	5	1	4	2

2.3.4.14 Conclusion to this Subsection

In this study, coils were produced using synthetic resins, PP and PE. The coils were mainly open-type coils without ferrite or capacitors. The electrical and mechanical properties of the coils were evaluated before and after burying them. Next, the mechanical strength of the road before and after the burial of the coils was evaluated using the FWD test. Four construction methods were investigated. Strain gauges were used to measure the deformation inside the coils, and thermocouples were used to measure and evaluate the effect of the heat of the asphalt mixture on the space inside the coils during paving. The efficiencies of Cases 2 and 4 were low owing to the pronounced influence of cementitious materials. Considering the efficiency, power, and road durability, Case 1 with the RC mesh had the highest overall evaluation. Future work could include the design of coils for higher power, investigation of the cause of the decrease in efficiency after burial, residual strain, and increasing the bearing capacity of the road pavement.

ACKNOWLEDGMENT FOR THIS SUBSECTION

This research was partly carried out as "A study on coil burial for dynamic wireless power transfer," by the commissioned research of National Institute for Land and Infrastructure Management under technology research and development system of the Committee on Advanced Road Technology established by MLIT, Japan, and partly supported by JSPS KAKENHI Grant Number 17H04915.

2.3.5 WPT on Power Passage between Rebars

Wireless power transmission [45] by magnetic field resonance coupling has the problem that power loss occurs and transmission efficiency deteriorates when metallic foreign bodies enter between the coils. For locations where power needs to be transmitted through reinforced concrete walls that utilize steel bars, one of the metallic foreign bodies, it is necessary to consider transmission based on the assumption that the steel bars have already been inserted. Here, we present a study [45] that examined methods to improve efficiency by testing the differences in the effects of different transmission distances, rebar spacing, rebar insulation, and other conditions. The conclusion is that insulating the intersection of the rebars reduces the deviation of the resonance frequency and improves the efficiency.

2.3.5.1 Background to the Rebar WPT Study

Wireless power transmission by magnetic field coupling has a problem that heat is generated when metallic foreign objects enter between the transmission coils. Metallic foreign objects must be detected from the viewpoint of human body protection and system protection, and various methods have been studied and proposed [46].

On the other hand, for locations where power must be transmitted through walls, such as closed spaces composed of reinforced concrete, power must be transmitted with foreign objects such as metals in the system. In other words, it is necessary to take an approach that considers transmission in a state in which a metallic foreign body such as a steel bar has entered, rather than an approach that detects or stops the system against a metallic foreign body, which has been studied so far.

Studies of power transmission in the presence of metallic foreign objects in electrical installations [47] and power transmission evaluation in the presence of reinforced concrete [48] have been done. In addition, earlier studies have clarified the effect of rebar on efficiency as the frequency increases, but they have not shown a method for improving efficiency. Therefore, in this study, different approaches such as different transmission distances, rebar spacing, and rebars with insulation treatment were taken to improve the transmission efficiency at specific resonance frequencies.

2.3.5.2 Coil and Rebar

A helical coil wound with Litz wire (number of strands: 240; diameter: 0.1 mm) as shown in Figure 2.80 was fabricated for the measurement. The coil was wound on an acrylic pipe with a diameter of 500 mm and thickness of 50 mm. The number of turns is 15 for both the transmitter and receiver coils.

The rebar between the transmitting and receiving coils was placed in a grid pattern as shown in Figure 2.81, using D16 rebar (each rebar has a diameter of 16 mm). In this case, the distance between the rebars is 100 mm, and the total size of the rebars is 1.5 m × 1.5 m square. The intersections of the rebars are bound with steel wire.

2.3.5.3 Four Different Conditions

The resonant frequency is 100 kHz. The test is in the form shown in Figures 2.82 and 2.83, and the following different conditions are used to propose appropriate efficiency improvement methods:

FIGURE 2.80   Coil fabricated.

FIGURE 2.81   Rebar.

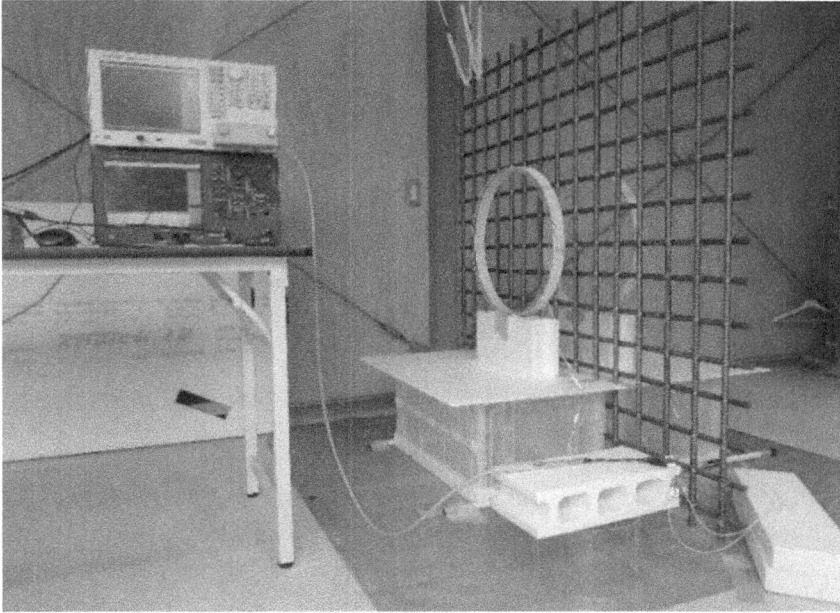

FIGURE 2.82   Transmission distance 250 mm, rebar spacing 100 mm.

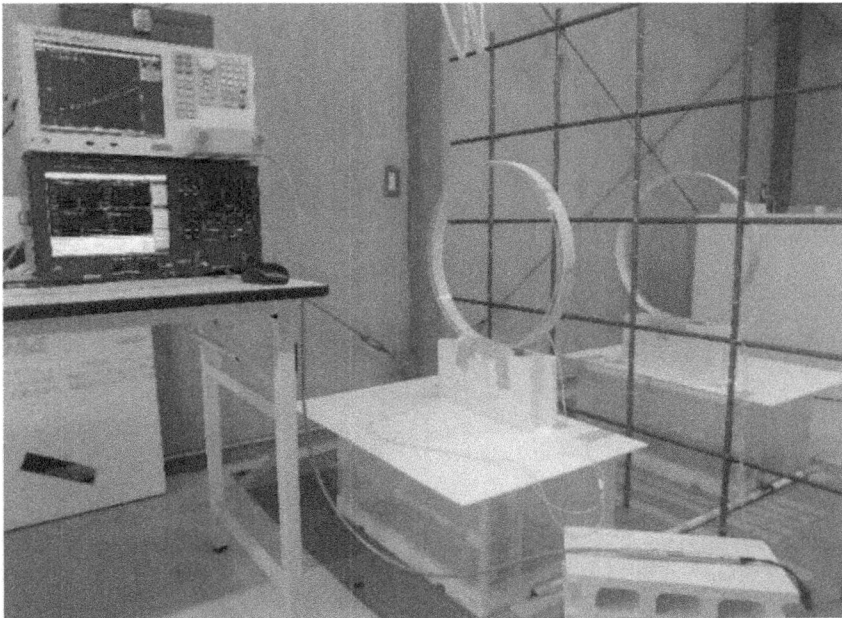

FIGURE 2.83   Transmission distance 700 mm, rebar spacing 300 mm.

- Transmission distance 250, 375, 500, 625, 750 mm

- Rebar spacing 100 m, 200 m, 300 mm

- Resonance frequency shift

- Insulation of rebar

Normally, the closer the transmission distance, the higher the transmission efficiency, but under conditions where there is a rebar in between, the effect may be reduced by moving away from the rebar.

Increasing the distance between rebars is expected to improve efficiency because the amount of metal is reduced. However, considering the original purpose of rebars used for reinforcement, it is considered undesirable to increase the spacing excessively.

The rebar causes a shift in resonance frequency. Correcting this misalignment will enable transmission at the highest efficiency.

The crossing section of the rebar is insulated so that eddy currents flowing through it may be stopped at the insulated section. This is expected to reduce the eddy current loss.

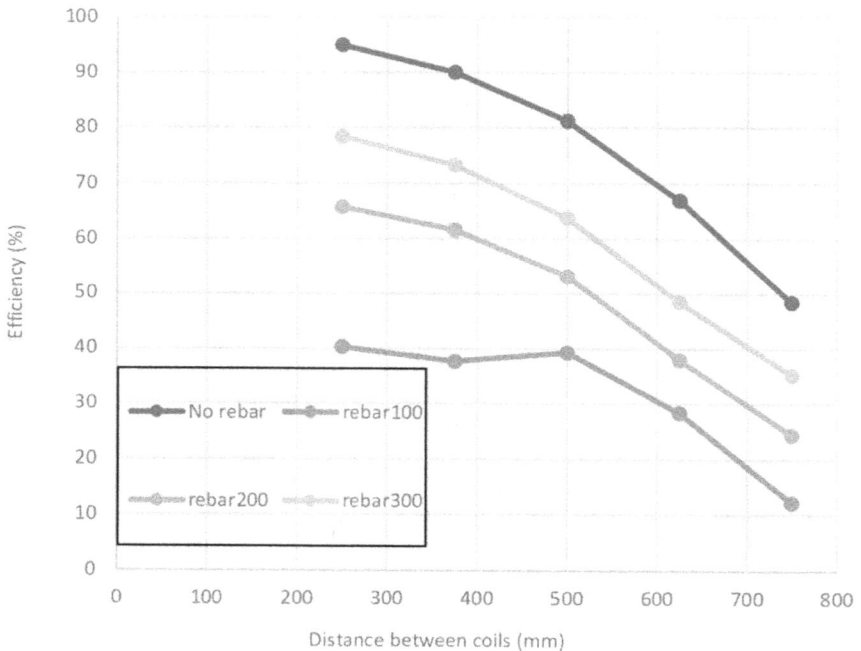

FIGURE 2.84 Efficiency at different inter-transmission distances and rebar spacing.

We expect that the transmission efficiency will be improved under these different conditions.

2.3.5.3.1 Experiment: Transmission Distance and Rebar Spacing   Measurements were made using a vector network analyzer (VNA). First, the results of tests in which the transmission distance and rebar spacing were varied are shown in Figure 2.84.

It was predicted that the efficiency would be improved by separating the transmission distance from the rebar, thereby reducing the influence of the rebar. The results of the test showed the predicted trend, although it did not lead to a result that could be called an improvement.

In particular, there was almost no change in efficiency between 250 and 500 mm for the 100 mm rebar. This is due to the strong influence of the rebar in the areas close to the transmission distance, where the resonance frequency is shifted.

2.3.5.3.2 Experiment: Correction of Resonance Frequency Shift   Next, the efficiency when the resonance frequency shift is corrected is shown in Figure 2.85.

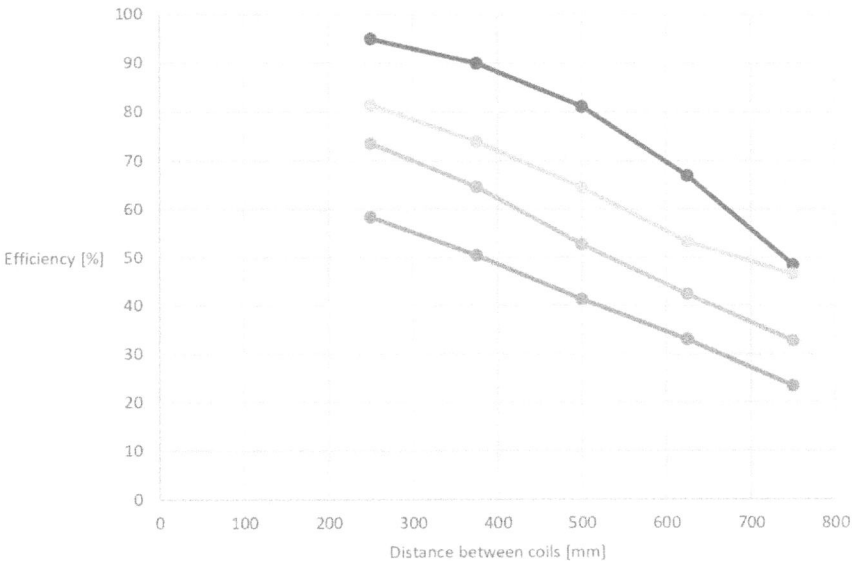

FIGURE 2.85   Efficiency when the resonance frequency shift is corrected.

By aligning the frequencies that were shifted by the rebar, it was expected that the efficiency would improve at short distances, where the effect of the rebar is more pronounced.

As predicted, the efficiency was improved at short distances, including a 20% increase at a distance of 250 mm between coils and 100 mm rebar spacing.

FIGURE 2.86   Insulation of rebar with tape.

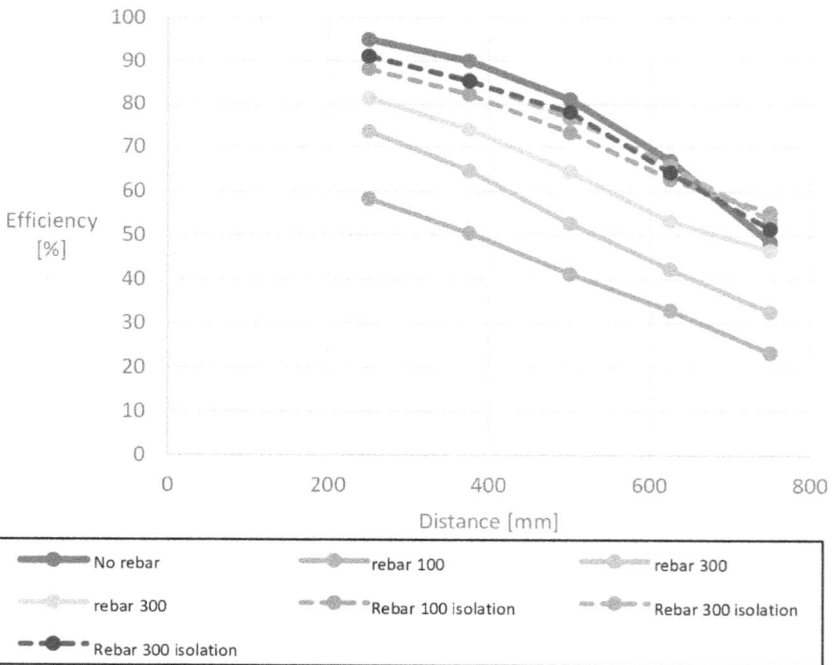

FIGURE 2.87   Efficiency of insulated rebar.

2.3.5.4 Experiment: Rebar Insulation

Finally, a transmission test with the rebar insulated is described. As shown in Figure 2.86, the rebar was insulated at the intersection of the rebars using tape.

Figure 2.87 compares the results of the test with the rebar insulated with the contents of Figure 2.85. When the rebars are insulated, there is almost no difference in efficiency whether the distance between the rebars is 100 mm, 200 mm, or 300 mm, and the efficiency is 90% at a transmission distance of 250 mm, a significant improvement over the 60% efficiency under the same conditions without insulation.

Figure 2.88 shows a graph of efficiency with frequency on the horizontal axis for the cases with and without insulation of the rebar at a transmission distance of 250 mm. As shown in the graph, the coil was originally fabricated with a resonance frequency of 100 kHz, but the presence of

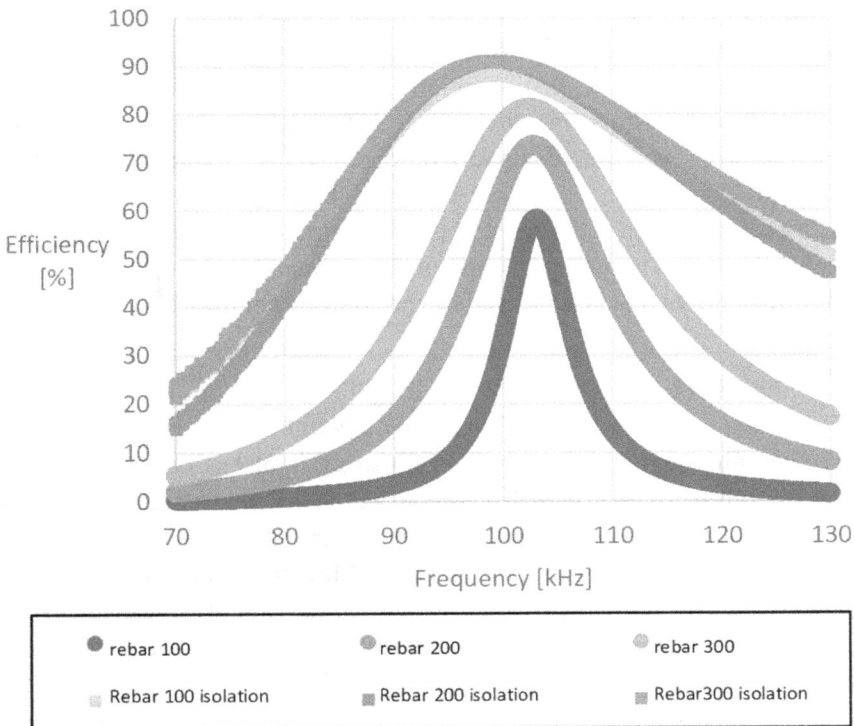

FIGURE 2.88 Improved rebar insulation and frequency misalignment.

Source: [45].

uninsulated steel bars between the bars causes a shift in the resonance frequency. On the other hand, when the rebar is insulated, not only is the efficiency greatly improved, but the resonance frequency shift is also improved, and it can be seen that the coil resonates at 100 kHz. In other words, the method of insulating the steel bar intersections can improve not only the efficiency but also the resonance frequency deviation.

2.3.5.5 Conclusion to the Rebar WPT

The influence of rebar on the transmission efficiency of wireless power transmission by magnetic field resonance coupling was examined using different approaches, such as transmission distance, rebar spacing, correction of resonance frequency deviation, and whether the rebar is insulated or not. The results showed that the frequency deviation was suppressed by reducing the influence of the rebar when the transmission distance was increased, and that the efficiency did not decrease even when the transmission distance was increased under the condition that the rebar was densely packed. It was also shown that if the resonance frequency deviation is corrected, the efficiency improves to 60% from 40% before the correction.

On the other hand, insulating the steel bars at their intersections improved the transmission efficiency to 90% under the same conditions, and also improved the deviation of the resonant frequency that was observed when the bars were not insulated. Based on these test results, it can be concluded that the most appropriate method to improve the efficiency of wireless power transmission by resonant coupling of magnetic fields through steel bars is to insulate the crossings of the bars.

2.3.6 Inducing Cancer Cell Death by WPT

2.3.6.1 Outline of Inducing Cancer Cell Death by WPT

Recently, photodynamic therapy (PDT) using light has been anticipated as a low-invasive treatment because it has fewer side effects than conventional cancer treatment [49]. However, current PDT is limited to the types of cancer that can be reached by cables, so it is difficult to apply PDT to cancers deep inside the body, such as pancreatic cancer. Here, we designed a wireless power transfer system that combines a small and lightweight receiving coil and LEDs for implantation in the body. After that, we experimented to see whether cancer cell death was induced by photoirradiation in vitro using power transfer in free air space. As a result, we succeeded in inducing about 75% cancer cell death within 60 minutes of photoirradiation at a 30 mm transfer distance. At that time, we derived an approximation of the amount of light energy required to induce cancer cell death by changing the photoirradiation time, and were able to predict the treatment time.

2.3.6.2 Introduction to Inducing Cancer Cell Death by WPT

Recently, WPT, which transfers power to various devices without cables, has been studied widely [50]. Through the establishment of technology using resonance, it enables higher efficiency and power transfer with a large air gap than conventional WPT, so it has been investigated for application in a wide range of fields, from electric vehicles [51–53] to smartphones [54,55]. Among these, WPT for implantable devices such as the ventricular assist devices [56,57] and the cochlear implants [58] has attracted much attention as a method to enhance the quality of life of patients due to its low invasiveness [59–61]. This technology eliminates the risk of surgery to replace batteries used in implantable devices in the body, infections caused by cables penetrating the skin, and battery breakage in the body because it becomes unnecessary to replace batteries or to supply power via cables from an external power source.

Here, we focus on the application to the medical field and target the integration of WPT and photodynamic therapy, a type of cancer treatment, at 6.78 MHz allocated in the ISM band. We designed a small and lightweight WPT device that can emit visible light using LEDs and evaluated it mainly in terms of radiant flux. After that, we verified its significance through the induction of cancer cell death in vitro.

2.3.6.3 Photodynamic Therapy

2.3.6.3.1 About Photodynamic Therapy Various treatments for cancer, the leading cause of death globally and among the three leading causes of death in Japan, are being researched internationally. Currently, surgery, chemotherapy, and radiation therapy are the three main established methods of cancer treatment. However, each method of cancer treatment has its own problems, such as physical burden, side effects, metastasis, and recurrence, and various new methods of cancer treatment, such as immunotherapy and targeted therapy, are being researched.

Here, we focus on one of these treatments, photodynamic therapy (PDT). PDT is a method of cancer treatment in which a photosensitizer that accumulates specifically on cancer cells is administered, followed by irradiation of cancer cells with light of a specific wavelength. As shown in the PDT mechanism of action in Figure 2.89, cancer cell death is induced

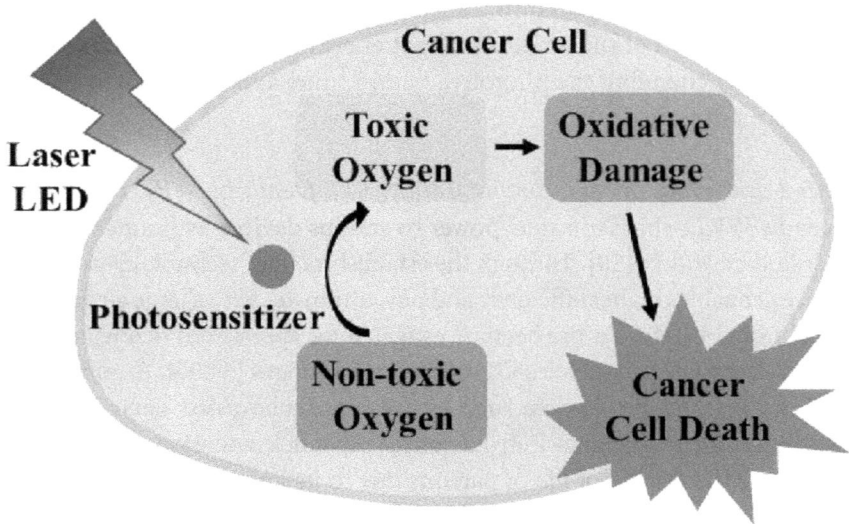

FIGURE 2.89   PDT mechanism of action.

Source: [49].

by generating singlet oxygen, a type of reactive oxygen species (ROS) that poses cytotoxicity within cancer cells. This treatment has less adverse effects on normal tissues and fewer side effects than existing treatments. In addition, it is said to induce damage or occlusion of nutrient blood vessels around cancer, thereby having an anti-tumor effect through a secondary action. In Japan, insurance coverage is applicable to lung cancer, esophageal cancer, gastric cancer, and malignant brain tumors using two types of photosensitizers: first-generation porfimer sodium and second-generation talaporfin sodium. However, current PDT is limited to applicable cancers, because laser irradiation is mainly performed by a combination of endoscopes and fiber-optic cables. Therefore, it is difficult to apply to cancers deep inside the body, such as pancreatic cancer.

2.3.6.3.2 Advantages of WPT+PDT Although photoirradiation of cancer with wire described in the previous section is basically completed in a single treatment, multiple photoirradiations are desirable for more effective treatment [62]. Therefore, implanting a light source for treatment in the body has been proposed, but there are only a few studies integrating WPT and PDT [63–65]. In addition, the transfer distance in those studies was short, and their application to cancers deep inside the body has not been considered. Here, we intend to integrate WPT

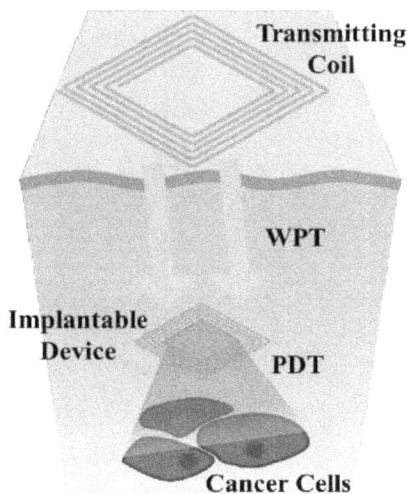

FIGURE 2.90   Image of cancer treatment with WPT+PDT.

Source: [66].

and PDT, as shown in Figure 2.90, and verify their significance by basic experiments using cancer cells in vitro. By doing so, we can contribute to future cancer treatment through the possibility of minimally invasive approaches to cancers deep inside the body, which are difficult to apply with current methods of cancer treatment. In addition, multidisciplinary treatment combined with other treatment methods including surgery reduces the recurrence risk caused by residual tissue, and elimination of the need for repeated surgery provides flexible treatment at recurrence. Using wirelessly PDT, which was previously performed by wire, there has been a number of advantages.

2.3.6.4 WPT System Design

Radiation type using microwaves or ultrasonic waves [67,68] and coupling type using magnetic fields or electric fields [69,70] are the main methods of WPT, however, here, we focus on WPT using magnetic fields. The wireless power transfer circuit to be used is a series (S)–parallel (P) topology magnetic field resonance circuit, in which capacitors and inductors are connected in series on the transmitting side and connected in parallel on the receiving side to resonate. The equivalent circuit diagram is shown in Figure 2.91. The fabricated transmitting and receiving devices are shown in Figure 2.92, and their respective specifications are shown in Table 2.14.

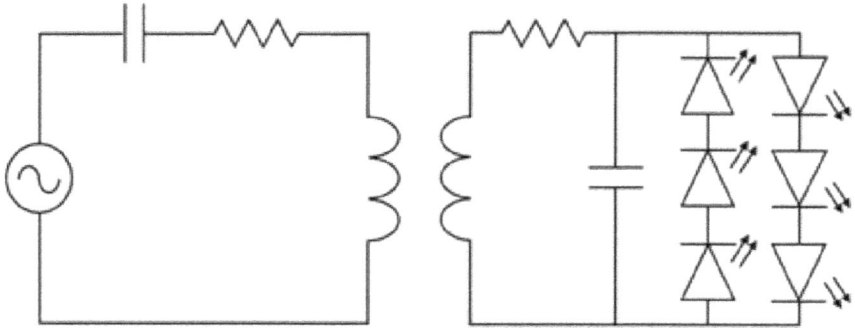

FIGURE 2.91   Equivalent circuit of WPT system (S–P).

Source: [49].

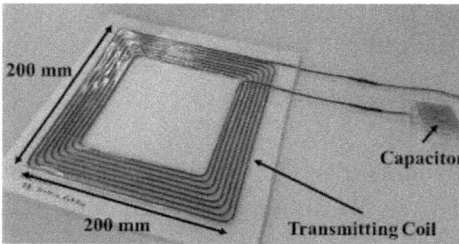

(a) Transmitting device   (b) Receiving device

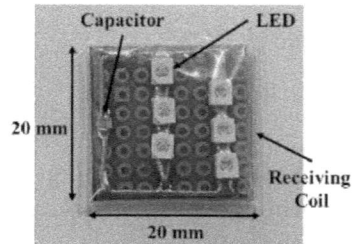

FIGURE 2.92   Pictures of transmitting and receiving devices.

TABLE 2.14   Specifications of Experimental Equipment

	Transmitting side	Receiving side
Size (mm)	200 × 200	20 × 20
Number of turns	8	6
Q value	449	83.0
Resistance (Ω)	2.35	0.55
Inductance (μH)	24.6	1.07
Capacitance (pF)	23.5	508

2.3.6.5 Evaluation of the WPT System

2.3.6.5.1 Measurement of Radiant Flux from LEDs
The radiant flux from the LEDs that emit blue light on the receiving device was measured using the experimental equipment shown in Figure 2.93. The operating frequency is set to 6.78 MHz, the input voltage is 8–10 V, and the transfer

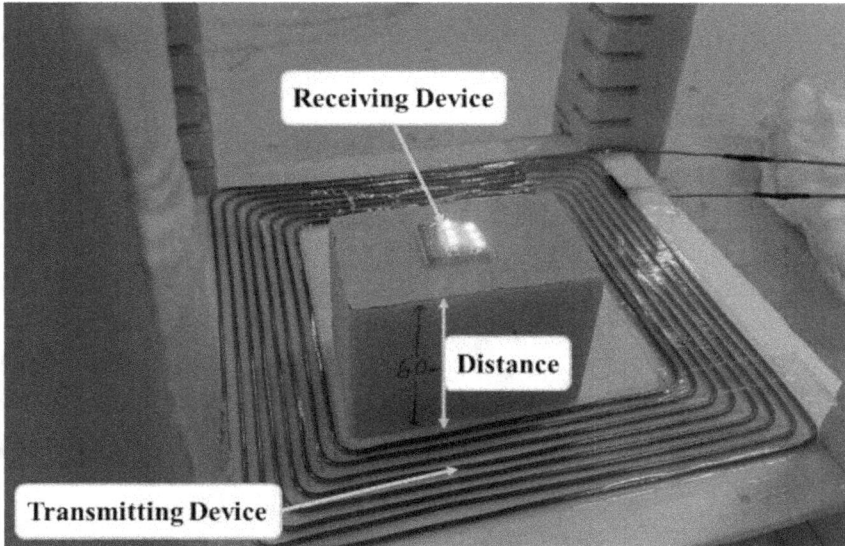

FIGURE 2.93   Measurement of radiant flux from LEDs.

Source: [49].

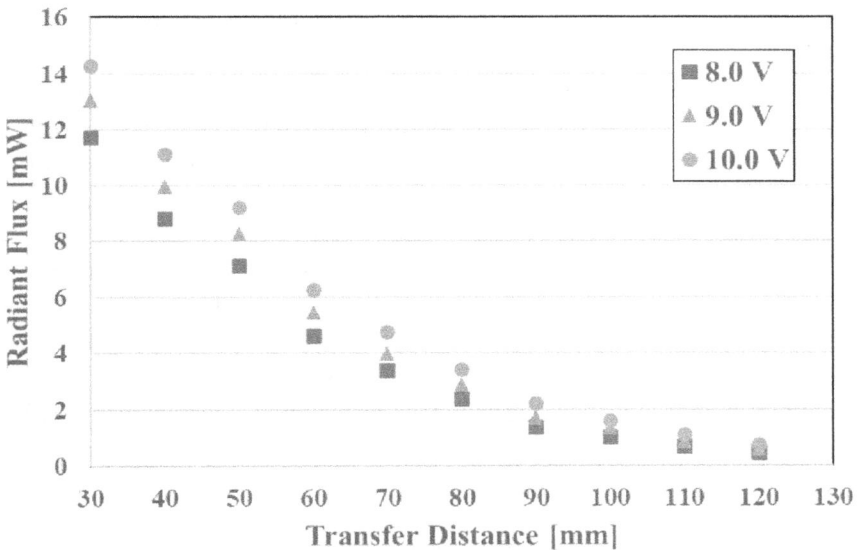

FIGURE 2.94   Measurement of the variation of the radiant flux from the LEDs at different input voltages (8–10 V) with different transfer distances (30–120 mm).

Source: [49].

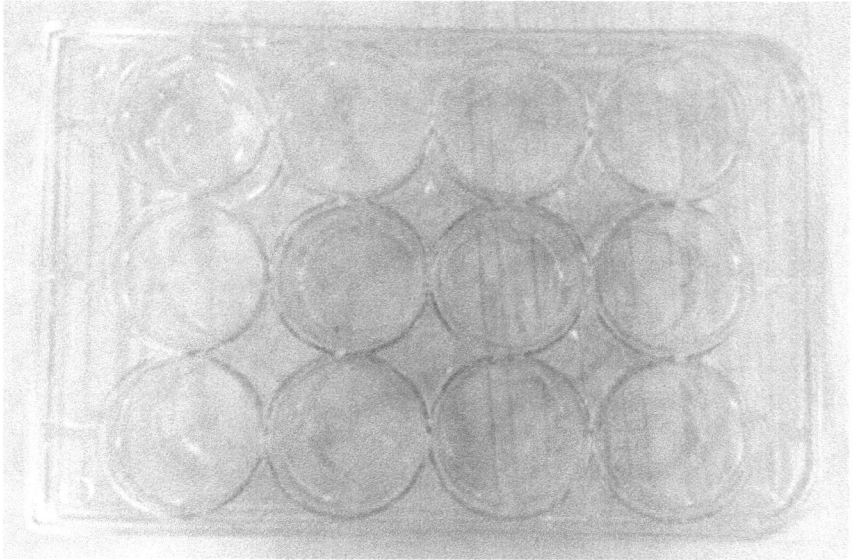

FIGURE 2.95   Culture plates for cell experiments.

Source: [49].

distance is 30–120 mm. When measuring the radiant flux, we considered the 15 mm space that exists between the cell surface to be cultured in the well plate and the emitting surface of the LED. As shown in Figure 2.94, the radiant flux from the LEDs on the receiving device was increased as the input voltage increased, and decreased as the transfer distance increased [71].

2.3.6.5.2 Distribution Evaluation of Radiant Flux from LEDs We evaluated whether sufficient photoirradiation was provided to the cell surface to be cultured when conducting cell experiments with the WPT device. The well plates in which cancer cells are cultured in cell experiments is shown in Figure 2.95. The standard for this well plate is cylindrical, with a depth of 15 mm and diameter of 22 mm.

2.3.6.6 Inducing Cancer Cell Death
Inducing cancer cell death was performed using the WPT system shown in Figure 2.96. Cancer cell death induced by singlet oxygen generated by photoirradiation of WPT using human uterine cervical cancer cells (HeLa-S3) and Ir(III) complexes was evaluated using nucleic acid staining dye (Propidium iodide: PI). PI does not permeate the cell membrane of living

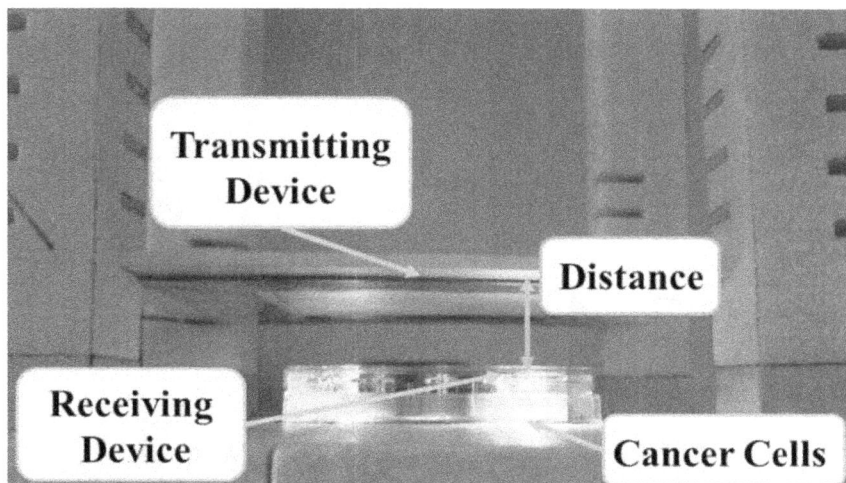

FIGURE 2.96　Inducing cancer cell death.

Source: [49].

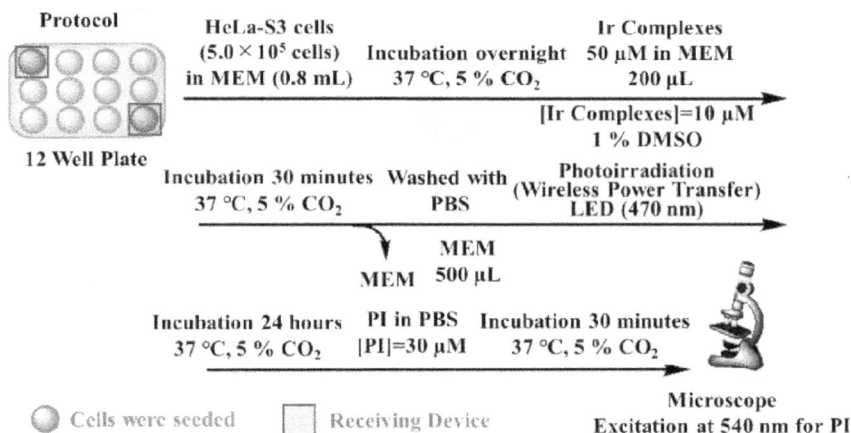

FIGURE 2.97　The protocol for inducing cancer cell death.

Source: [49].

cells, but migrates into the nucleus of dead cells and emits a remarkable red fluorescence as it is incorporated into the DNA. The protocol for inducing cancer cell death is shown in Figure 2.97.

Fluorescence microscopy images of HeLa-S3 cells are shown in Figure 2.98 under the reaction conditions in Table 2.15 with an input voltage of 10 V, a transfer distance of 30 mm, and an incubation time of

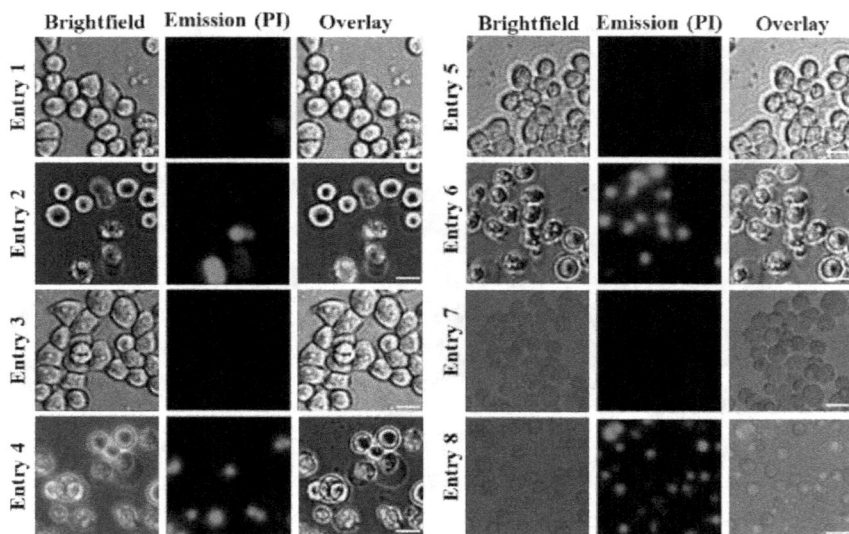

FIGURE 2.98   Fluorescence microscopic image of HeLa-s3 cell. Entry 1: minutes photo irradiation, without Ir(III) complexes, Entry 2: 30 minutes photoirradiation, Ir(III) complexes, Entry 3: 40 minutes photoirradiation, without Ir(III) complexes, Entry 4: 40 minutes photoirradiation, Ir(III)complexes, Entry 5: 50 minutes photoirradiation without Ir(III)complexes, Entry 6: 50 minutes photoirradiation, Ir(III) complexes), Ir(III)complexes, Entry 7: 60 minutes photoirradiation, without Ir(III) complexes, Entry 8:60 minutes photoirradiation, without Ir(III) complexes. The white scale bar is 20 μ m.

TABLE 2.15   Reaction Conditions

Entry	Photosensitizer	Transfer distance (mm)	Photoirradiation time (min)
1	—	30	30
2	Ir(III) complexes	30	30
3	—	30	40
4	Ir(III) complexes	30	40
5	—	30	50
6	Ir(III) complexes	30	50
7	—	30	60
8	Ir(III) complexes	30	60

24 hours after photoirradiation. The photoirradiation times were 30, 40, 50, and 60 minutes. In Entry 1, Entry 3, Entry 5, and Entry 7, where Ir(III) complexes were absent, there was little or no induction of cancer cell death by photoirradiation. On the other hand, in Entry 2, Entry 4, Entry 6, and Entry 8, where Ir(III) complexes were present, the cancer cells were clearly

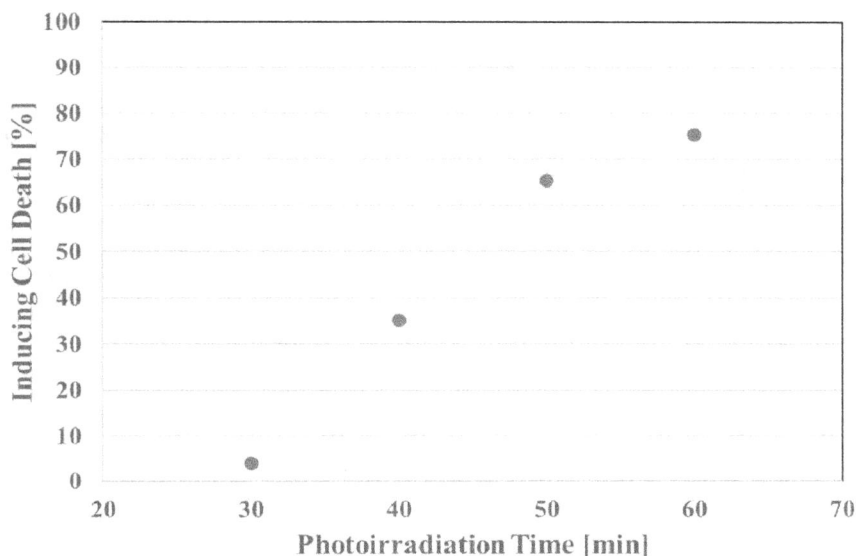

FIGURE 2.99   Percentage of induced cancer cell death.

Source: [49].

stained red and fluorescent. These results indicate that cancer cell death was induced by singlet oxygen generated by photoirradiation.

Next, we investigated the effects of photoirradiation time on cancer cell death. As shown in Figure 2.99, the number of cancer cells stained red increased as the photoirradiation time was increased to 30, 40, 50, and 60 minutes, which means that the percentage of inducing cell death increased to 4%, 35%, 65%, and 75%.

2.3.6.7 Conclusion to Inducing Cancer Cell Death by WPT

Here, we proposed methods of cancer treatment that integrate WPT and PDT with 6.78 MHz as the operating frequency and verified its significance by circuit simulations and experiments. A WPT system that combined a small and lightweight receiving coil and LEDs for implantation in the body was designed based on circuit simulations, and evaluated in terms of engineering and pharmaceuticals. After engineering evaluation based on two characteristics of LEDs, current value and radiant flux, and pharmaceutical evaluation based on singlet oxygen generation by drug reaction experiments using DPBF, this WPT system was confirmed in advance to have sufficient performance for inducing cancer cell death. After that, cancer cell death was induced in vitro as a basic experiment for cancer treatment. Under the condition of 30 mm transfer distance in free space, about 75% cancer cell death was induced successfully after 60 minutes of photoirradiation. These results demonstrated the expanded possibilities for cancer treatment by integrating WPT and PDT.

2.3.7 Seawater and WPT

WPT in water and WPT in seawater are completely different [72]. Wireless power transfer in mere water is easily successful, but in seawater it is very difficult. This subsection introduces WPT research in seawater.

2.3.7.1 Outline of Seawater and WPT

Wireless power transfer in seawater has been a problem due to seawater being a highly conductive medium causing eddy current losses to occur between the transmitter and receiver, making it difficult to obtain stable power efficiency. Therefore, we added a polyethylene sheet between the coil and seawater and verified the improvement of efficiency by increasing the spacer distance. We also revealed the frequency characteristics of the coils depending on the separation distance. These are shown through multiple experiments.

2.3.7.2 Introduction

Japan has the world's sixth largest Exclusive Economic Zone, and the existence of many resources such as oil, natural gas, methane hydrate, and seafloor hydrothermal deposits have been confirmed in the surrounding waters. In order to exploit and supply these resources, it is essential to construct an efficient seafloor mapping and exploration system. This is why autonomous underwater vehicles (AUVs) have been developed in recent years. Currently, AUVs have limited operating time due to the limitations of their internal batteries. In addition, when recharging the battery, the AUVs must be brought up to the water's surface and lifted by crane ships, which requires time and effort. As a solution to this problem, wireless power transfer in seawater has been proposed [73–75]. As shown in Figure 2.100, a power supply spot is set up in the seawater and AUVs can actively supply power for long hours of operation. However, because seawater is highly conductive, eddy current losses occur in seawater, making it difficult to obtain stable power efficiency [76,77]. Since the optimal frequency band

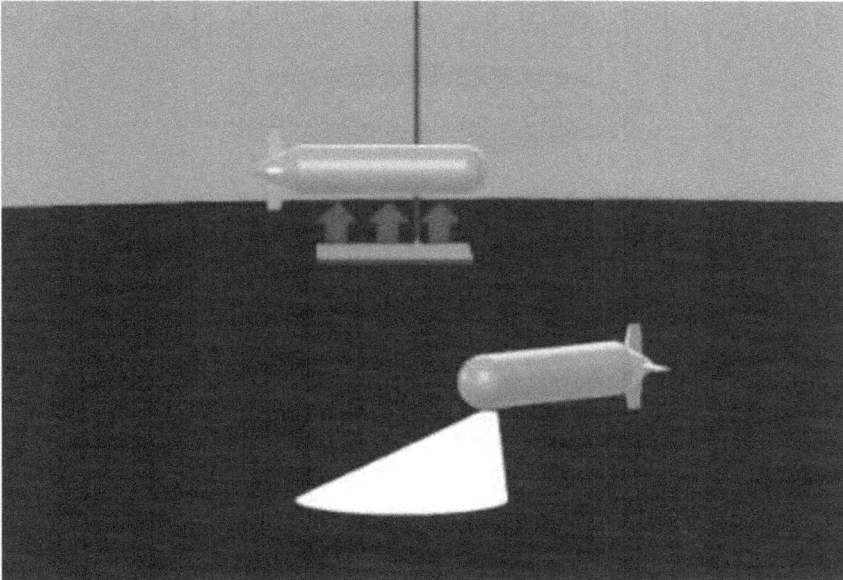

FIGURE 2.100   Image of underwater wireless (power transfer).

for wireless power transfer in seawater is unclear, changes in frequency characteristics due to it will be revealed here. We will also compare and reveal the characteristics from the analysis and the experiment, when spacers are placed between the coil and seawater.

2.3.7.3 Change in Coil Characteristics due to Seawater

2.3.7.3.1 Equivalent Circuit of Coil with Eddy Current Loss   Seawater is a medium with high conductivity, and the presence of seawater near the coil causes eddy currents in the seawater as shown in Figure 2.101, which deteriorates the coil characteristics. A diagram showing the circuit for Figure 2.101 is shown in Figure 2.102(a), and the T-type equivalent circuit of Figure 2.102(a) is shown in Figure 2.102(b) [78].

From Kirchhoff's law, Equation (2.56) is obtained.

$$\begin{pmatrix} V \\ 0 \end{pmatrix} = \begin{pmatrix} j\omega(L_{coil} + L_m) + R_{coil} & -j\omega L_m \\ -j\omega L_m & j\omega(L_{eddy} + L_m) + R_{eddy} \end{pmatrix} \begin{pmatrix} I_{coil} \\ I_{eddy} \end{pmatrix} \quad (2.56)$$

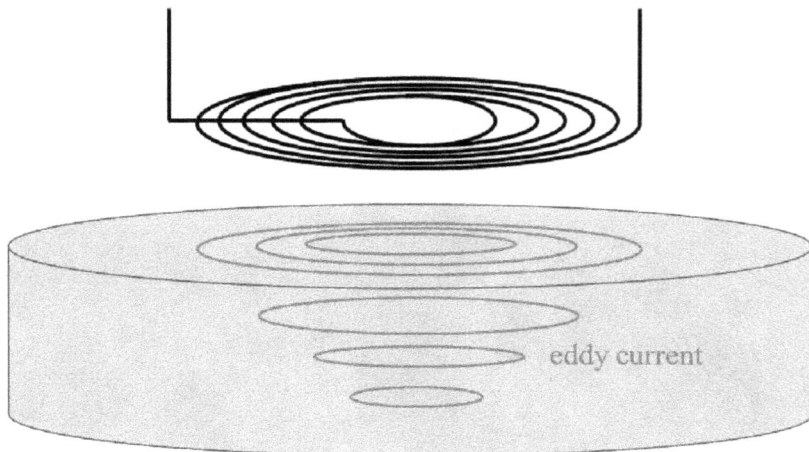

FIGURE 2.101 Model of coil with eddy current.

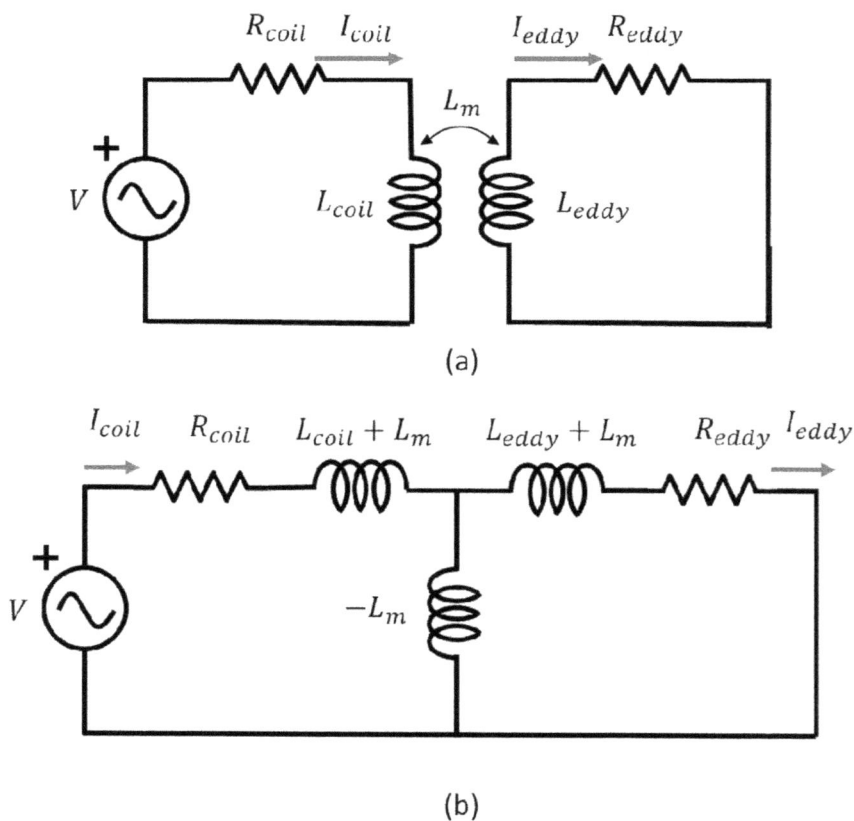

(a)

(b)

FIGURE 2.102 Equivalent circuit of coil with eddy current loss.

Next, the synthetic impedance Z_{all} is derived and shown in Equation (2.57).

$$Z_{all} = R_{coil} + j\omega L_{coil}$$

$$+ \frac{\omega^2 L_m^2 R_{eddy} - j\omega \left\{ \omega^2 L_{eddy} \left(L_{eddy} + L_m \right) + R_{eddy}^2 \right\}}{R_{eddy}^2 + \omega^2 L_{eddy}^2} \qquad (2.57)$$

From Equation (2.57), the equivalent internal resistance R_{all} of the coil in seawater is expressed as in Equation (2.58).

$$R_{all} = R_{coil} + \frac{\omega^2 L_m^2 R_{eddy}}{R_{eddy}^2 + \omega^2 L_{eddy}^2} \qquad (2.58)$$

From Equation (2.58), the equivalent internal resistance R_{all} of the coil in seawater is expressed as a function of ω. From Equations (2.57) and (2.58), the Q value of the coil in seawater is shown in Equation (2.59).

$$Q = \frac{\omega \left\{ R_{eddy}^2 \left(L_{coil} - 1 \right) + \omega^2 L_{eddy} \left(L_{coil} L_{eddy} - L_{eddy} - L_m \right) \right\}}{R_{coil} \left(R_{eddy}^2 + \omega^2 L_m^2 \right) + \omega^2 L_m^2 + R_{eddy}} \qquad (2.59)$$

2.3.7.3.2 Verification by Simulation Next, the equivalent circuit and equations obtained in Section 2.1 are verified by analysis and experiment. First, the verification is performed by electromagnetic field analysis using the method of moments. The coil used in the analysis is 350 mm × 350 mm on the outside, with 10 turns and a wire diameter of 3 mm. The model used in the analysis is shown in Figure 2.103.

The seawater volume is 700 × 700 × 300 mm, the transmission distance is 100 mm, and a seawater model is created with a 0.2 mm space between the coil and seawater to prevent seawater from conducting current. Table 2.16 shows the parameters used in the simulation. ε_r is the relative permittivity, μ_r is the relative permeability, and σ is the conductivity. A graph of the frequency characteristics of the inductance L and coupling coefficient k

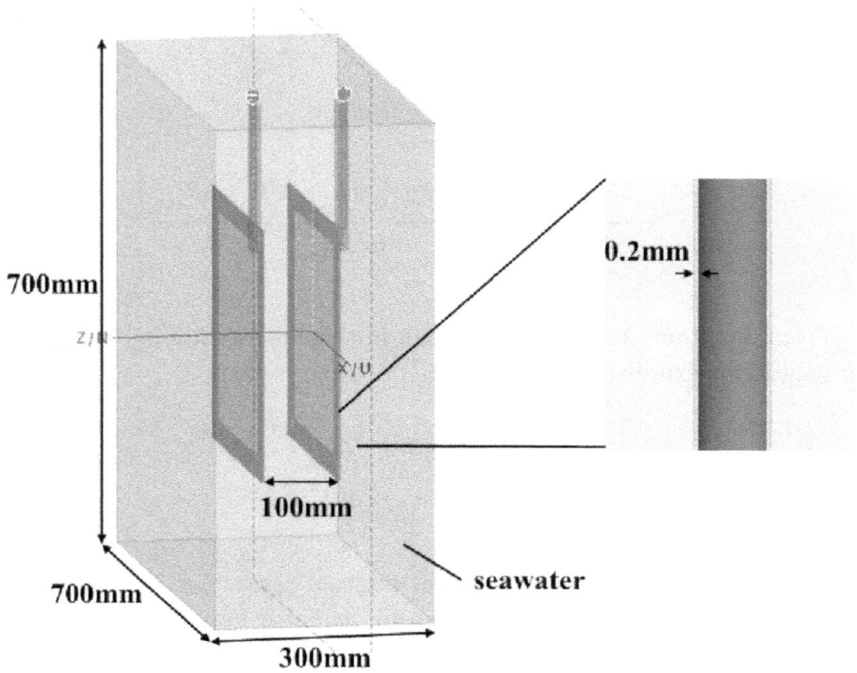

FIGURE 2.103   Simulation model for underwater coils.

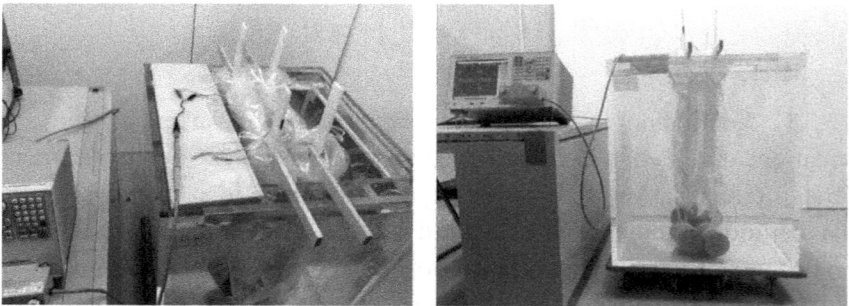

FIGURE 2.104   Measurement setup for underwater coils.

TABLE 2.16   Simulation Parameters

	ε_r	μ_r	σ (S/m)	Specific gravity [kg/m^3]
Seawater	78	1	5.3	1020

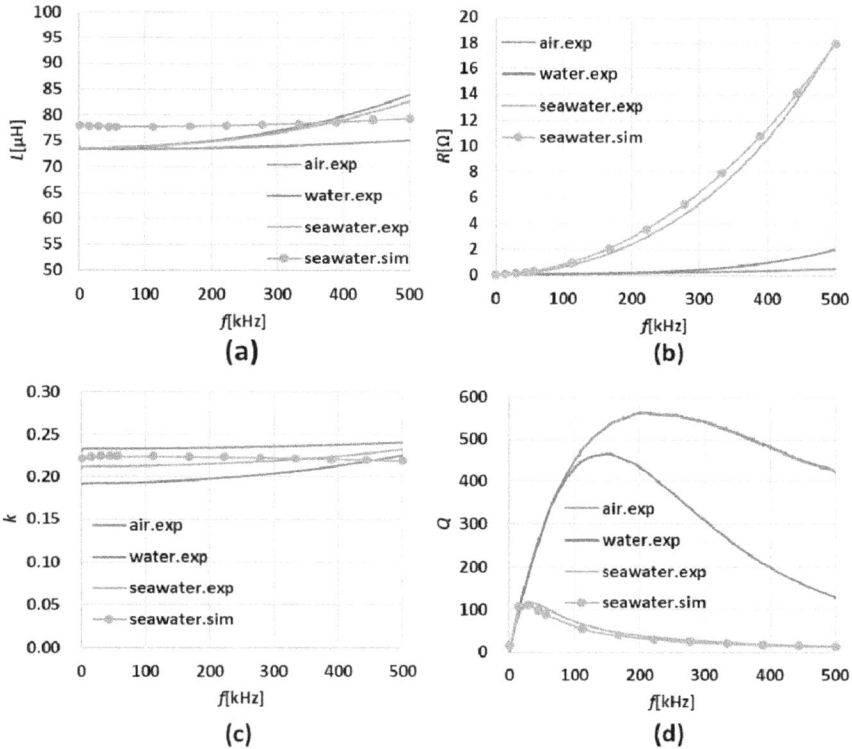

FIGURE 2.105   Simulation and measurement results of frequency characteristics of coils: (a) inductance, (b) resistance, (c) coupling coefficient, (d) Q.

of the coil in seawater obtained from the analysis is shown in Figure 2.105; Figure 2.106 shows a graph of the maximum efficiency η_{max} derived from the analytical results using Equation (2.60) [79]. The measured values of the graphs are shown in subsection 2.3.7.3.3.

$$\eta_{max} = \frac{k^2 Q^2}{\left(1 + \sqrt{1 + k^2 Q^2}\right)^2}. \qquad (2.60)$$

2.3.7.3.3 verification by Experiment   Next, the frequency characteristics of an actual tank (700 × 700 × 800 mm) filled with artificial seawater are measured. For the measurement, a coil of the same size as that used in the

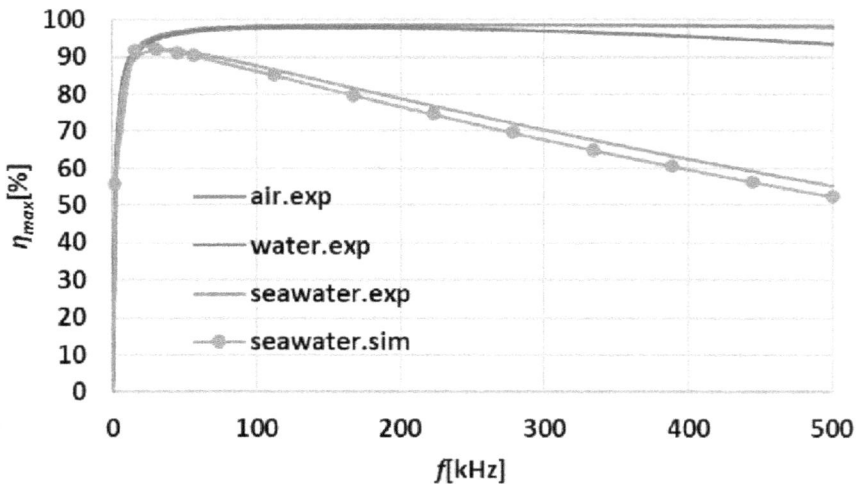

FIGURE 2.106 Simulation and measurement results of maximum frequency of underwater coils.

TABLE 2.17 Maximum Q Value of Coil

	f (kHz)	Q_{max}	η_{max} (%).
Air	208.8	565.0	99.1
Water	150.0	465.8	98.5
Seawater	34.1	115.0	93.6

analysis was created, and the measurement was performed using impedance analyzers. The actual experiment is shown in Figure 2.104.

The results of the frequency response of the inductance L and coupling coefficient k obtained from the measurement and analysis are shown in Figure 2.105. Figure 2.106 shows a graph of the maximum efficiency η_{max} derived from Equation (2.5) using the results obtained from Figure 2.105. The analytical and experimental results show that the internal resistance of the coil in seawater increases in proportion to the frequency, and the Q value decreases significantly due to this effect. As shown in Table 2.17, the maximum Q value Q_{max} was 565 at 208.8 kHz in air, while it was 115 at 34.1 kHz in seawater. The coupling coefficient does not change significantly, and

Figure 2.106 shows that the maximum efficiency η_{max} also decreases as the Q value decreases.

2.3.7.4 Change in Characteristics with Spacer

2.3.7.4.1 Verification by Simulation   Next, the change in coil characteristics when spacing is added between the coils and seawater between the transmitter and receiver is verified by analysis and experiment (Figure 2.107). The coil geometry and model used in the analysis are the same as in the previous section. From there, free space is placed between the coil and seawater. The width of the spacer is s and analysis is performed for $s = 0, 5, 10, 20, 30, 40,$ and 50 mm, respectively, for comparison. At $s = 50$ mm, all the space between the transmitter and receiver is free space. The results of the analysis are shown in Figures 2.108 and 2.109.

2.3.7.4.2 Verification by Experiment   The following are the results of the experiment. Foamed polyethylene was attached as a spacer to the coil, and

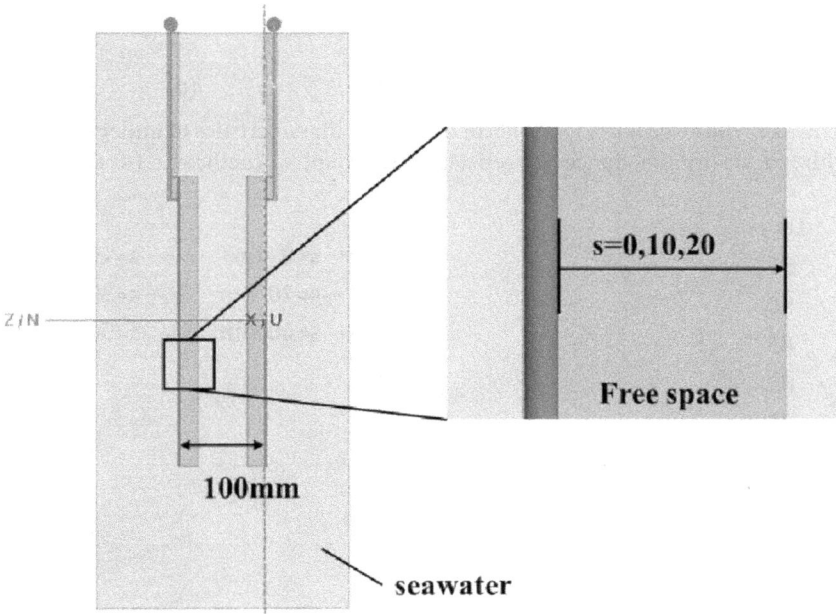

FIGURE 2.107   Simulation model for underwater coils with inserting spacer between seawater and coil.

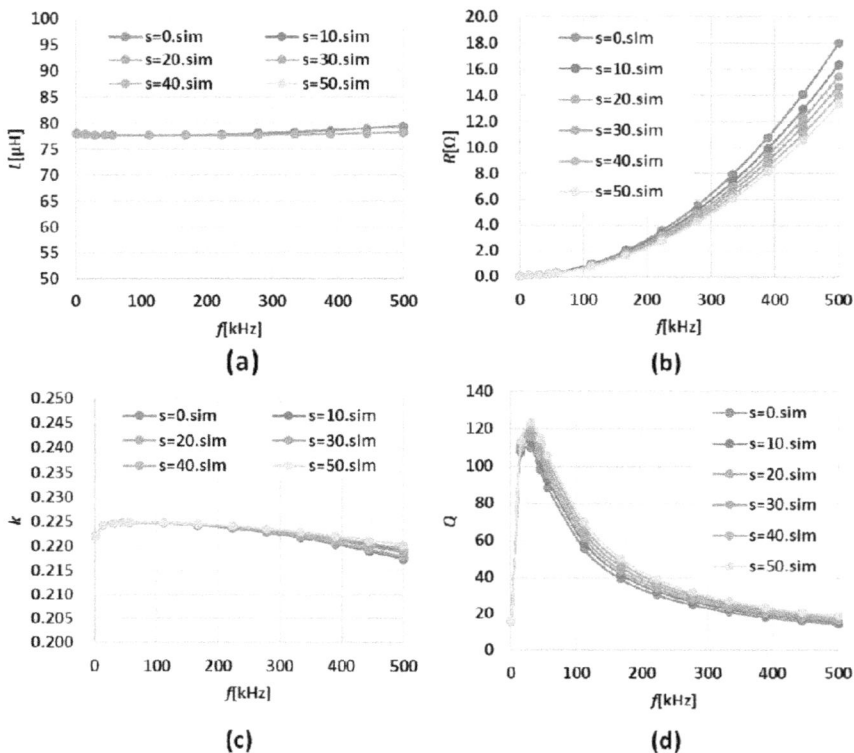

FIGURE 2.108   Simulation results of frequency characteristics of underwater coils with spacer: (a) inductance, (b) resistance, (c) coupling coefficient, (d) Q.

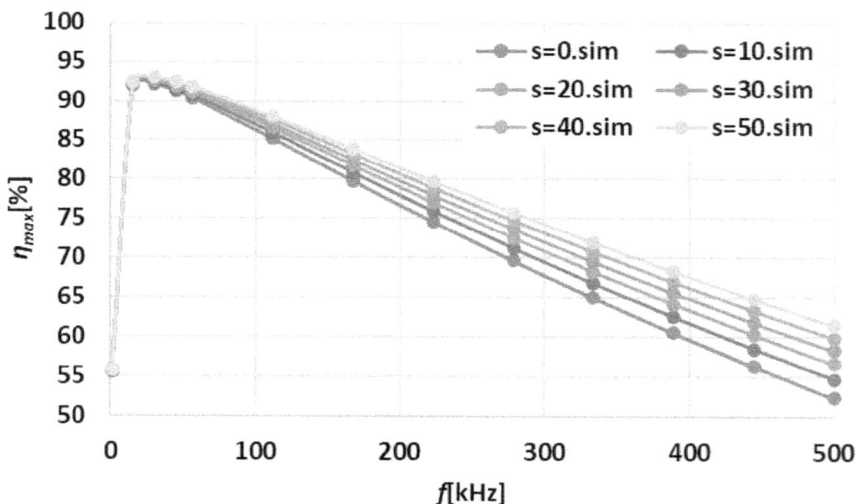

FIGURE 2.109   Simulation results of maximum frequency of underwater coils with spacer.

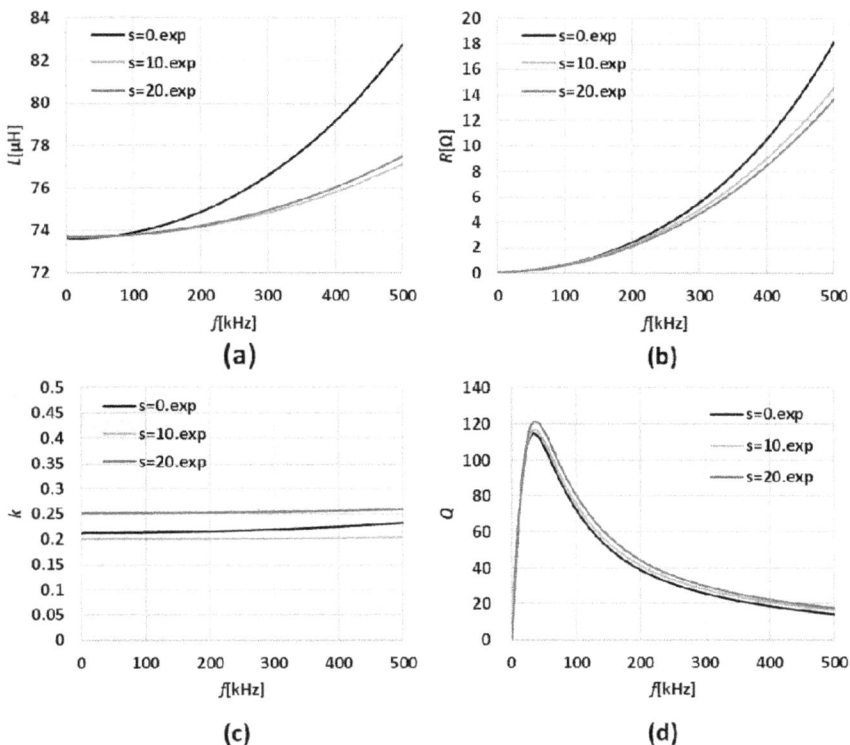

FIGURE 2.110 Measurement results of frequency characteristics of underwater coils with spacer: (a) inductance, (b) resistance, (c) coupling coefficient, (d) Q.

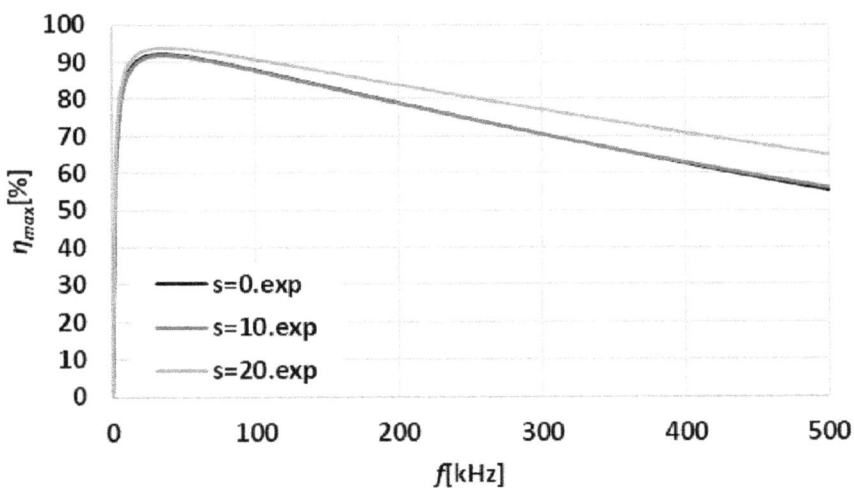

FIGURE 2.111 Measurement results of maximum frequency of underwater coils with spacer.

measurements were taken at $s = 0, 5,$ and $10\,mm$, respectively, as in the analysis. The measured results are shown in Figures 2.110 and 2.111.

Figures 2.110 and 2.111 show that inductance L and internal resistance R vary with the spacer, and that the larger the width of spacer s, the higher the Q value of the coil. However, there is almost no change in the coupling coefficient k, even when there is no seawater between the transmitter and receiver, indicating that the main cause of the lower efficiency of wireless power transfer in seawater is the deterioration of the characteristics of the coil itself rather than the attenuation of the magnetic field in seawater. The maximum efficiency η_{max} increased by 1.3% for $s = 0$ mm and $s = 50$ mm at $45\,kHz$, while it increased by 9.1% at $500\,kHz$, indicating that the improvement in coil characteristics by the spacer is greater at higher frequencies.

2.3.7.5 Conclusion

Analyses and experiments were conducted on the characteristics of coils in air, water, and seawater media, respectively, to compare and reveal the changes in frequency characteristics. It can be seen that the internal resistance of the coil in seawater increases in proportion to the frequency, and the Q value decreases significantly due to this effect. Although the characteristics of the coil used in this study are also relevant, it had a Q_{max} of 115 at 34.1 kHz in seawater, resulting in a maximum efficiency of 93.6%. It was also found that the coupling coefficient did not change significantly and the maximum efficiency η_{max} also decreased as the Q value decreased. Next, the changes in coil characteristics by inserting spacers ($s = 0, 10, 20$ mm) between the coil and seawater were compared and verified through analyses and experiments. It was found that the main reason for the efficiency decreasing in seawater was the deterioration of coil characteristics, and that the spacers improved the Q value of the coil, with the higher frequency having a greater effect with the spacers.

2.4 CHAPTER SUMMARY

In this chapter, we have introduced magnetic field coupling, mainly magnetic field resonant coupling (magnetic field resonance), which was considered a mystery at the start of 2007, but as we have shown, it is a method that makes very good use of the resonance phenomenon of electromagnetic induction. Resonant circuit configurations from S–S to LCC were also explained. As application examples, wireless power supply to stationary EVs, control and burial technology for dynamic WPT, WPT passing through steel bars, cancer treatment and WPT, and seawater and WPT were also introduced.

REFERENCES

[1] G. A. Covic and J. T. Boys, "Inductive Power Transfer," in *Proceedings of the IEEE*, vol. 101, no. 6, pp. 1276–1289, June 2013.

[2] Green, A. W. and Boys, J. T., "An Inductively Coupled High Frequency Power System for Material Handling Applications." in *Proceedings of the IPEC Conference*, vol. 2, Singapore, pp. 18–19, 821–826, March 1993.

[3] André Kurs, Aristeidis Karalis, Robert Moffatt, J. D. Joannopoulos, Peter Fisher and Marin Soljačić, "Wireless Power Transfer via Strongly Coupled Magnetic Resonances," *Science Express*, vol. 317, no. 5834, pp. 83–86, June 2007.

[4] Takehiro Imura, "*Wireless Power Transfer: Using Magnetic and Electric Resonance Coupling Techniques*," Springer Nature Singapore Pte. Ltd., 2020.

[5] O. H. Stielau and G. A. Covic, "Design of Loosely Coupled Inductive Power Transfer Systems," *Proceedings of the International Conference on Power System Technology*, vol. 1, pp. 85–90, 2000.

[6] Takehiro Imura and Yoichi Hori, "Unified Theory of Electromagnetic Induction and Magnetic Resonant Coupling," *IEEJ D*, vol. 135, no. 6, pp. 697–710, 2015.

[7] Takehiro Imura and Yoichi Hori, "Superiority of Magnetic Resonant Coupling at Large Air Gap in Wireless Power Transfer," *42nd Annual Conference of the IEEE Industrial Electronics Society*, IEEE, c2016, Oct. 2016.

[8] Yuto Yamada and Takehiro Imura, "An Efficiency Optimization Method of Static Wireless Power Transfer Coreless Coils for Electric Vehicles in the 85 kHz Band Using Numerical Analysis," *Ieej Transactions on Electrical and Electronic Engineering, Ieej Trans 2022*, vol. 17, no. 10, pp. 1506–1516, October 2022.

[9] Takehiro Imura, Hiroyuki Okabe, Toshiyuki Uchida and Yoichi Hori, "Study on Open and Short End Helical Antennas with Capacitor in Series of Wireless Power Transfer using Magnetic Resonant Couplings," *IEEE Industrial Electronics Society Annual Conference*, pp. 3848–3853, 2009.

[10] www.felicanetworks.co.jp/en/mfelica_pf/history.html   [Accessed: 18 Dec. 2022]

[11] Koichi Furusato, Takehiro Imura, and Yoichi Hori, "Design of Multi-frequency Coil for Capacitor-less Wireless Power Transfer using High Order Self-resonance of Open End Coil", *The IEEE MTT-S Wireless Power Transfer Conference*, 2016.

[12] Y. H. Sohn, B. H. Choi, E. S. Lee, G. C. Lim, G. Cho, and C. T. Rim, "General Unified Analyses of Two-Capacitor Inductive Power Transfer Systems: Equivalence of Current-Source SS and SP Compensations," *IEEE Transactions on Power Electronics*, vol. 30, no. 11, pp. 6030–6045, 2015.

[13] B. Wei et al., "*A Study on Constant Current Output Control in Wireless Power Transfer,*" pp. 2–5, 2017.

[14] F. Liu, Y. Zhang, K. Chen, Z. Zhao, and L. Yuan, "A Comparative Study of Load Characteristics of Resonance Types in Wireless Transmission Systems," *Asia-Pacific International Symposium on Electromagnetic Compatability APEMC 2016*, pp. 203–206, 2016.

[15] Kodai Takeda and Takafumi Koseki, "*Analytical Investigation on Asymmetric LCC Compensation Circuit for Trade-off between High Efficiency and Power,*" IPEC, 2018.

[16] Kanta Sasaki, Takehiro Imura, "Combination of Sensorless Energized Section Switching System and Double-LCC for DWPT," *2020 IEEE PELS Workshop on Emerging Technologies; Wireless Power (WoW2020)*, Nov. 2020. IEEE: Seoul.

[17] T. Kan, S. Member, T. Nguyen, J. C. White, R. K. Malhan, and C. C. Mi, "A New Integration Method for an Electric Vehicle Wireless Charging System Using LCC Compensation Topology: Analysis and Design," *IEEE Transactions on Power Electronics*, vol. 32, no. 2, pp. 1638–1650, 2017.

[18] Motoki Sato, G. Yamamoto, Daisuke Gunji, Takehiro Imura and Hiroshi Fujimoto, "Development of Wireless In-Wheel Motor using Magnetic Resonance Coupling," *Transactions on Power Electronics, IEEE*, vol. 31, no. 7, pp. 5270–5278, 2016.

[19] www.sae.org/standards/content/j2954_202208/ [Accessed: 18 Dec. 2022]

[20] D. Kobayashi, K. Hata, T. Imura, H. Fujimoto and Y. Hori, "Sensorless Vehicle Detection Using Voltage Pulses in Dynamic Wireless Power Transfer System," *29th International Electric Vehicle Symposium and Exhibition*, 2016.

[21] Kanta Sasaki, Takehiro Imura, "Combination of Sensorless Energized Section Switching System and Double-LCC for DWPT," *2020 IEEE PELS Workshop on Emerging Technologies; Wireless Power (WoW2020)*, IEEE, c2020, Nov. 2020.

[22] C. Mi, G. Buja, S. Y. Choi, and C. T. Rim, "Modern Advances in Wireless Power Transfer Systems for Roadway Powered Electric Vehicles," *IEEE Transaction Industrial Electronics*, vol. 63, no. 10, pp. 6533–6545, 2016.

[23] G. A. Covic and J. T. Boys, "Modern Trends in Inductive Power Transfer for Transportation Applications," *IEEE Journal of Emerging and Selected Topics in Power Electronics*, vol. 1, no. 1, pp. 28–41, 2013.

[24] A. Kurs, A. Karalis, R. Moffatt, J. D. Joannopoulos, P. Fisher, and M. Solja?i?, "Wireless Power Transfer Via Strongly Coupled Magnetic Resonances," *Science*, vol. 317, no. 5834, pp. 83–86, 2007.

[25] Y. Choi, Su, Beom W. Gu, Seog Y. Jeong, Chun T. Rim, "Advances in Wireless Power Transfer Systems for Roadway-Powered Electric Vehicles," *IEEE Journal of Emerging and Selected Topics in Power Electronics*, vol. 3, no. 1, pp. 18–36, 2014.

[26] K. Hata, K. Hanajiri, T. Imura, H. Fujimoto, Y. Hori, M. Sato, and D. Gunji, "Driving Test Evaluation of Sensorless Vehicle Detection Method for In-motion Wireless Power Transfer," *2018 International Power Electronics Conference (IPEC-Niigata 2018-ECCE Asia)*, pp. 663–668, 2018.

[27] C. Wang, P. Wang, Q. Zhu, and M. Su, "An Alternate Arrangement of Active and Repeater Coils for Quasi-Constant Power Wireless EV Charging," 2019 *IEEE PELS Workshop on Emerging Technologies: Wireless Power Transfer (WoW)*, 2019.

[28] K. Song, C. Zhu, K. E. Koh, D. Kobayashi, T. Imura, and Y. Hori, "Modeling and Design of Dynamic Wireless Power Transfer System for EV Applications," *IECON 2015-41st Annual Conference of the IEEE Industrial Electronics Society*, pp. 005229–005234, 2015.

[29] Takehiro Imura, Koki Hanawa, Kanta Sasaki and Nagato Abe, "Coil Performance and Evaluation of Pavement Durability of Dynamic Wireless Power Transfer System using Ferrite-less and Capacitor-less Coil for Road Construction Methods," *5th International Electric Vehicle Technology Conference (EVTeC2021)*, May 2021.

[30] G. A. Covic and J. T. Boys, "Modern Trends in Inductive Power Transfer for Transportation Applications," *IEEE Journal of Emerging and Selected Topics in Power Electronics*, vol. 1, no. 1, pp. 28–41, 2013.

[31] Su Y. Choi, Beom W. Gu, Seog Y. Jeong, Chun T. Rim, "Advances in Wireless Power Transfer Systems for Roadway-Powered Electric Vehicles," *IEEE Journal of Emerging and Selected Topics in Power Electronics*, vol. 3, no. 1, pp.18–36, 2014.

[32] K. Song, C. Zhu, K. E. Koh, D. Kobayashi, T. Imura, and Y. Hori, "Modeling and Design of Dynamic Wireless Power Transfer System for EV Applications," *IECON 2015-41st Annual Conference of the IEEE Industrial Electronics Society*, pp. 005229–005234, 2015.

[33] D. Kobayashi, K. Hata, T. Imura, H. Fujimoto, and Y. Hori, "Sensorless Vehicle Detection Using Voltage Pulses in Dynamic Wireless Power Transfer System," *29th International Electric Vehicle Symposium and Exhibition*, 2016.

[34] C. C. Mi, G. Buja, S. Y. Choi, and C. T. Rim, "Modern Advances in Wireless Power Transfer Systems for Roadway Powered Electric Vehicles," *IEEE Transactions on Industrial Electronics*, vol. 63, no. 10, pp. 6533–6545, 2016.

[35] K. Hata, K. Hanajiri, T. Imura, H. Fujimoto, Y. Hori, M. Sato, and D. Gunji, "Driving Test Evaluation of Sensorless Vehicle Detection Method for In-motion Wireless Power Transfer," *2018 International Power Electronics Conference (IPEC-Niigata 2018-ECCE Asia)*, pp. 663–668, 2018.

[36] C. Wang, P. Wang, Q. Zhu, and M. Su, "An Alternate Arrangement of Active and Repeater Coils for Quasi-Constant Power Wireless EV Charging," *2019 IEEE PELS Workshop on Emerging Technologies: Wireless Power Transfer (WoW)*, 2019.

[37] M. Maemura and A. Wendt, "Dynamic Power Transfer as a Feature – Employing Stationary WPT Devices for Dynamic Operation," *2020 IEEE PELS Workshop on Emerging Technologies: Wireless Power Transfer (WoW)* pp. 50–55, 2020.

[38] V. Z. Barsari, D. J. Thrimawithana, G. A. Covic, and S. Kim, "A Switchable Inductively Coupled Connector for IPT Roadway Applications," *IEEE Transactions on Power Electronics*, vol. 1, no. Icc, pp. 35–39, 2020.

[39] B. J. Varghese, A. Kamineni, N. Roberts, M. Halling, D. J. Thrimawithana, and R. A. Zane, "Design Considerations for 50 kW Dynamic Wireless Charging with Concrete-Embedded Coils," *IEEE PELS Workshop on Emerging Technologies: Wireless Power Transfer (WoW)*, pp. 40–44, 2020.

[40] R. M. Nimri, A. Kamineni, and R. Zane, "A Modular Pad Design compatible with SAE J2954 for Dynamic Inductive Power Transfer," *IEEE PELS Workshop on Emerging Technologies: Wireless Power Transfer (WoW)*, pp. 45–49, 2020.

[41] Koichi Furusato, Takehiro Imura and Yoichi Hori, "Improvement of 85 kHz Self-resonant Open End Coil for Capacitor-less Wireless Power Transfer System," *2016 Asian Wireless Power Transfer Workshop*, Dec. 2016.

[42] Yoshiaki Takahashi, Takehiro Imura and Yoichi Hori, "Comparison of 85 kHz Self-resonant Open-end Coils with Different Types of Wire for Capacitor-less Wireless Power Transfer System," *2017 Asian Wireless Power Transfer Workshop*, Dec. 2017.

[43] Yoshiaki Takahashi, Katsuhiro Hata, Takehiro Imura and Yoichi Hori, "Comparison of Capacitor- and Ferrite-less 85kHz Self-Resonant Coils Considering Dielectric Loss for In-motion Wireless Power Transfer," *The 44th Annual Conference of the IEEE Industrial Electronics Society*, Oct. 2018.

[44] Saburo Matsuno, Taisuke Kobayashi, "About the Vehicle Driving Position," *The 14th Japan Road Conference*, pp. 177–178, 1981 (in Japanese).

[45] Seiya Yamamoto, et al., "A Method for Improving Efficiency in Wireless Power Transfer by Magnetic Resonant Coupling through Rebar," *IEICE, WPT2021-28*, copyright(c)2022 IEICE.

[46] Y. Sun, G. Wei, K. Qian, P. He, C. Zhu and K. Song, "A Foreign Object Detection Method Based on Variation of Quality Factor of Detection Coil at Multi-frequency," *2021 IEEE 12th Energy Conversion Congress & Exposition – Asia (ECCE-Asia)*, pp. 1578–1582, 2021.

[47] H. W. R. Liang, H. Wang, C. -K. Lee and S. Y. R. Hui, "Analysis and Performance Enhancement of Wireless Power Transfer Systems With Intended Metallic Objects," *IEEE Transactions on Power Electronics*, vol. 36, no. 2, pp. 1388–1398, 2021.

[48] Seung-Hwan Lee, Myung-Yong Kim, Byung-Song Lee and Jaehong Lee, "Impact of Rebar and Concrete on Power Dissipation of Wireless Power Transfer Systems," *IEEE Transactions on Industrial Electronics*, vol. 67, no. 1, pp. 276–287, Jan. 2020.

[49] Yoshitaka Yasuda, et al., "Studies on Inducing Cancer Cell Death and Required Light Energy by Photodynamic Therapy Using Wireless Power Transfer," *IEEE, SPEC2022 c2022 IEEE*.

[50] A. Kurs, A. Karalis, R. Moffatt, J. D. Joannopoulos, P. Fisher, and M. Soljačić, "Wireless Power Transfer via Strongly Coupled Magnetic Resonances," *Science*, vol. 317, no. 5834, pp. 83–86, July 2007.

[51] S. Li and C. C. Mi, "Wireless Power Transfer for Electric Vehicle Applications," *IEEE Journal of Emerging and Selected Topics in Power Electronics*, vol. 3, no. 1, pp. 4–17, March 2015.

[52] R. Tavakoli and Z. Pantic, "Analysis, Design, and Demonstration of a 25-kW Dynamic Wireless Charging System for Roadway Electric Vehicles," *IEEE Journal of Emerging and Selected Topics in Power Electronics*, vol. 6, no. 3, pp. 1378–1393, Sept. 2018.

[53] K. Sasaki and T. Imura, "Combination of Sensorless Energized Section Switching System and Double-LCC for DWPT," *2020 IEEE PELS Workshop on Emerging Technologies: Wireless Power Transfer (WoW)*, pp. 62–67, Nov. 2020.

[54] S. Nguyen, C. Duong and R. Amirtharajah, "A Smart Health Tracking Ring Powered by Wireless Power Transfer," *2021 IEEE Wireless Power Transfer Conference (WPTC)*, pp. 1–4, 2021.

[55] S. -M. Kim, I. -K. Cho, S. -W. Kim, J. -I. Moon and H. -J. Lee, "A Qi-compatible Wireless Charging Pocket for Smartphone," *2020 IEEE Wireless Power Transfer Conference (WPTC)*, pp. 387–390, 2020.

[56] Y. Liu, Y. Li, J. Zhang, S. Dong, C. Cui and C. Zhu, "Design a Wireless Power Transfer System with Variable Gap Applied to Left Ventricular Assist Devices," *2018 IEEE PELS Workshop on Emerging Technologies: Wireless Power Transfer (Wow)*, pp. 1–5, 2018.

[57] T. Campi, S. Cruciani, F. Maradei, A. Montalto, F. Musumeci and M. Feliziani, "Centralized High Power Supply System for Implanted Medical Devices Using Wireless Power Transfer Technology," in *IEEE Transactions on Medical Robotics and Bionics*, vol. 3, no. 4, pp. 992–1001, Nov. 2021.

[58] S. Hong et al., "Cochlear Implant Wireless Power Transfer System Design for High Efficiency and Link Gain Stability Using A Proposed Stagger Tuning Method," *2020 IEEE Wireless Power Transfer Conference (WPTC)*, pp. 26–29, 2020.

[59] D. Mukherjee and D. Mallick, "Experimental Demonstration of Miniaturized Magnetoelectric Wireless Power Transfer System For Implantable Medical Devices," *2022 IEEE 35th International Conference on Micro Electro Mechanical Systems Conference (MEMS)*, pp. 636–639, 2022.

[60] S. G. Jang, J. Kim, J. Lee, J. S. Kim, D. Hwan Kim and S. M. Park, "Wireless Power Transfer Based Implantable Neurostimulator," *2020 IEEE Wireless Power Transfer Conference (WPTC)*, pp. 365–368, 2020.

[61] Y. Ma, Z. Luo, C. Steiger, G. Traverso, and F. Adib, "Enabling Deep-Tissue Networking for Miniature Medical Devices," *Proceedings of ACM SIGCOMM 2018 Conference*, pp. 417–431, Aug. 2018.

[62] Mitsunaga M, Nakajima T, Sano K, Choyke PL, Kobayashi H, "Near-infrared Theranostic Photoimmunotherapy (PIT): Repeated Exposure of Light Enhances the Effect of Immunoconjugate," *Bioconjugate Chemistry*, vol. 23, no. 3, pp. 604–409, 2012.

[63] P. M. Lee, X. Tian and J. S. Ho, "Wireless Power Transfer for Glioblastoma Photodynamic Therapy," *2019 IEEE Biomedical Circuits and Systems Conference (BioCAS)*, pp. 1–4, 2019.

[64] Kirino, I., Fujita, K., Sakanoue, K. et al., "Metronomic Photodynamic Therapy Using an Implantable LED Device and Orally Administered 5-Aminolevulinic Acid," *Scientific Reports*, vol. 10, Dec. 2020.

[65] Kim, A., Zhou, J., Samaddar, S. et al., "An Implantable Ultrasonically-Powered Micro-Light-Source (µLight) for Photodynamic Therapy", *Scientific Reports*, vol. 9, Feb. 2019.

[66] Y. Yoshitaka, T. Imura, Y. Hori, K. Yokoi, A. Kanbe, M. Kakihana and S. Aoki, "A Basic Experiment with Cancer Cells for Photodynamic Therapy by Wireless Power Transfer," *IEICE Technical Report*, vol. 121, no. 289, WPT2021-14, pp. 5–8, Dec. 2021.

[67] B. Strassner and K. Chang, "Microwave Power Transmission: Historical Milestones and System Components," *Proceedings of the IEEE*, vol. 101, no. 6, pp. 1379–1396, June 2013.

[68] M. Meng and M. Kiani, "Design and Optimization of Ultrasonic Wireless Power Transmission Links for Millimeter-Sized Biomedical Implants," *IEEE Transactions on Biomedical Circuits and Systems*, vol. 11, no. 1, pp. 98–107, Feb. 2017.

[69] T. Imura and Y. Hori, "Maximizing Air Gap and Efficiency of Magnetic Resonant Coupling for Wireless Power Transfer Using Equivalent Circuit and Neumann Formula," *IEEE Transactions on Industrial Electronics*, vol. 58, no. 10, pp. 4746–4752, Oct. 2011.

[70] M. P. Theodoridis, "Effective Capacitive Power Transfer," *IEEE Transactions on Power Electronics,* vol. 27, no. 12, pp. 4906–4913, Dec. 2012.

[71] A. Kando, Y. Hisamatsu, H. Ohwada, T. Itoh, S. Moromizato, M. Kohno, and S. Aoki, "Photochemical Properties of Red-Emitting

Tris(cyclometalated) Iridium(III) Complexes Having Basic and Nitro Groups and Application to pH Sensing and Photoinduced Cell Death," *Inorganic Chemistry*, vol. 54, no. 11, pp. 5342–5357, Jun 2015.

[72] Yuki Miyakozawa, Takehiro Imura, Yoichi Hori, "Frequency Characteristics of Wireless Power Transfer in Seawater via Magnetic Resonant Coupling,", AWPT 2022.

[73] R. Hasaba, K. Eguchi, S. Yamaguchi, H. Satoh, T. Yagi and Y. Koyanagi, "WPT System in Seawater for AUVs with kW-class Power, High Positional Freedom, and High Efficiency inside the Transfer Coils," *2022 Wireless Power Week (WPW)*, pp. 90–94, 2022.

[74] J. Zhou, P. Yao, Y. Chen, K. Guo, S. Hu and H. Sun, "Design Considerations for a Self-Latching Coupling Structure of Inductive Power Transfer for Autonomous Underwater Vehicle," *IEEE Transactions on Industry Applications*, vol. 57, no. 1, pp. 580–587, Jan.-Feb. 2021.

[75] T. Kan, R. Mai, P. P. Mercier and C. C. Mi, "Design and Analysis of a Three-Phase Wireless Charging System for Lightweight Autonomous Underwater Vehicles," *IEEE Transactions on Power Electronics*, vol. 33, no. 8, pp. 6622–6632, Aug. 2018.

[76] Z. Liu, L. Wang, Y. Guo and C. Tao, "Eddy Current Loss Analysis of Wireless Power Transfer System for Autonomous Underwater Vehicles," *2020 IEEE PELS Workshop on Emerging Technologies: Wireless Power Transfer (WoW)*, pp. 283–287, 2020.

[77] K. Zhang, Y. Ma, Z. Yan, Z. Di, B. Song and A. P. Hu, "Eddy Current Loss and Detuning Effect of Seawater on Wireless Power Transfer," *IEEE Journal of Emerging and Selected Topics in Power Electronics*, vol. 8, no. 1, pp. 909–917, March 2020.

[78] Khokle, Rajas Prakash, Karu P. Esselle, and Desmond J. Bokor, "Design, Modeling, and Evaluation of the Eddy Current Sensor Deeply Implanted in the Human Body," *Sensors*, vol. 18, no. 11, p. 3888, 2018.

[79] K. Hata, T. Imura and Y. Hori, "Simplified measuring method of kQ product for wireless power transfer via magnetic resonance coupling based on input impedance measurement," *IECON 2017 – 43rd Annual Conference of the IEEE Industrial Electronics Society*, pp. 6974–6979, 2017.

CHAPTER 3

Wireless Power Transfer by Capacitive Coupling

Alessandra Costanzo

3.1 INTRODUCTION

Intentional wireless energy and information transfer, from a RF source to single or multiple users, located in the near-field of the source, is made possible by means of reactive electromagnetic energy exchange.

In such conditions, the electric $\mathbf{E}$ and magnetic $\mathbf{H}$ fields are in quadrature and the Poynting vector $\mathbf{E} \times \mathbf{H}$ flux is positive for the first half period, and represents outgoing energy flow, and is negative for the second half period, and represents inward energy flow.

Such a reactive interaction may be implemented by either exploiting magnetic induction, namely inductive power transfer (IPT), or electric induction, namely capacitive power transfer (CPT). To obtain sufficient efficiency, these two reactive mechanisms should be used when the transmitter and the receiver are in close proximity, which is at a distance of the order of a wavelength or lower.

IPT and CPT principles of operations are physically related to each other due to the duality principle and they theoretically offer the same capabilities.

However, despite the first WPT demonstration being based on capacitive coupling given by Tesla in 1891 [1,2], inductive WPT has been more

DOI: 10.1201/9781003328636-3

widely experimented on and more commercial products are already in the market, with respect to its dual implementation.

Focusing of the research on IPT rather than on CPT probably has been because CPT has been more widely employed for small gap distances between the transmitting and receiving devices and for low-power applications, such as consumer electronics [3], electrical machines, and biomedical devices [4], due to the difficulty of achieving a high coupling capacitance or high operating frequency.

However, with the increase in semiconductor power devices and the possibility to design a tunable compensating network, for both the receiver and transmitter sides, CPT systems are becoming attractive for long-distance high-power applications, especially for electric vehicle wireless charging, in such a way that they may be a competitive alternative to IPT [5–7]. In particular, CPT systems show a superior ability at reducing EMI shielding requirements, by exploiting the more directive nature of electric fields and greater robustness with respect to transmitter and receiver misalignments [8].

This chapter aims at providing the basic theoretical and numerical characterizations of the wireless front-end of a capacitive link.

After analyzing the main building blocks of a CPT architecture, and the related operating principles, circuit-equivalent models are derived and the main performance indexes are computed, which allow to rigorously predict the link performance, and to establish roles for the design of optimum terminations for both the CPT link source and the load. Some application examples taken from the most recent literature are then reported on.

3.2 MODEL OF THE WIRELESS CPT LINK

3.2.1 The End-to-End CPT Link

Figure 3.1 shows the block representation of an end-to-end CPT link, from the DC input source to the DC output load. The main building blocks are: the DC-to-RF converter with the matching/compensating network; the capacitive wireless link, schematically represented by two-faced metal plates; the RF-to-DC converter with the matching/compensating network; and the load which is the equivalent of the end-user subsystem, namely the battery or device.

The associated power quantities at each block interface are outlined in the same figure. By accurately predicting these quantities, the end-to-end

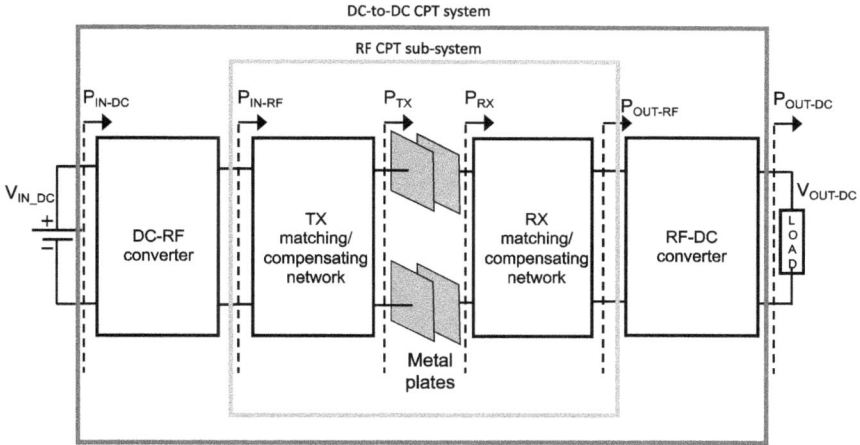

FIGURE 3.1 Schematic block representation of the end-to-end capacitive CPT link.

link efficiency η_{CPT} may be computed, at the CTP link operating frequency as [9]:

$$\eta_{CPT} = \eta_{DC-RF} \cdot \eta_{RF-RF} \cdot \eta_{RF-DC} \qquad (3.1)$$

The first term:

$$\eta_{DC-RF} = \frac{P_{TX}}{P_{IN-DC}}$$

is the ratio between the RF power, P_{TX}, at the wireless front-end input, and the DC power, P_{IN-DC}, required to support the DC-to-RF power conversion: this contribution is mainly due to the conversion efficiency of the RF power source (inverter or oscillator and amplifier).

The second term:

$$\eta_{RF-RF} = \frac{P_{RX}}{P_{TX}}$$

is the efficiency of the wireless link covered during the energy transfer operation (ranging from a few millimeters to a few centimeters in the case of resonant near-field WPT [NF-WPT], and a few centimeters to tens of centimeters in the cases of non-resonant NF-WPT and mid-field WPT [MF-WPT]). It consists of the ratio between the RF power, P_{TX}, at the wireless front-end input, and the output power at the output metal plate(s) P_{RX}. This quantity is highly dependent on the geometrical parameters of the wireless link, including the metal planes distance and misalignment and the surrounding materials.

Finally, the last term:

$$\eta_{RF-DC} = \frac{P_{OUT-DC}}{P_{RX}}$$

is the ratio between the dc power delivered to the final user, P_{OUT-DC}, and the received one, P_{RF}, i.e. the conversion efficiency of the rectifying section from the higher frequency to dc. It is typically given by the further product of two efficiencies: the $RF - DC$ conversion efficiency of the sole rectifier and the $DC - DC$ efficiency of the power management unit, which is adopted to load the rectifier at its optimum value. This value is established by the specifications associated with the link design, as will be discussed in Section 3.2.2.

It is worth mentioning that the WPT system consists of a connection of nonlinear circuits; for this reason, Equation (3.1) greatly depends on the involved power levels and operating frequencies: and can be accurately evaluated only if all phenomena (both linear and nonlinear) are rigorously taken into consideration, thus providing a precise estimate of the powers contributing to Equation (3.1). This is particularly true for the NF-WPT systems since both the TX and RX operating regimes strongly influence each other.

Indeed, the network resulting from the connection of the wireless link with the receiving sub-system is the actual load of the transmitting sub-system. Similarly, the network resulting from the connection of the transmitting sub-system and the wireless link is the actual source excitation of the receiving sub-system. Thus, variation of either the signal levels or the link distance or the metal plates misalignment affect the entire CPT link performance.

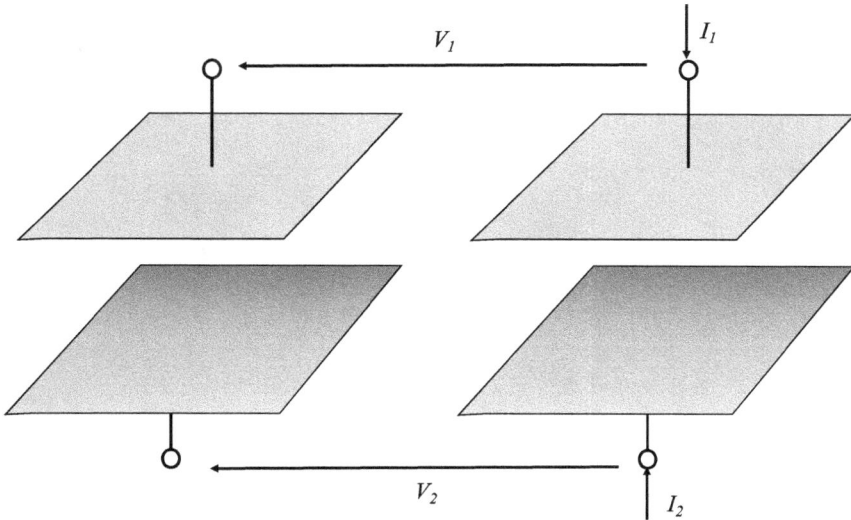

FIGURE 3.2   Perspective view of a horizontally aligned four-plate CPT.

3.2.2 Architectures of the Capacitive Wireless Link

Different arrangements for the wireless CPT links have been proposed for the coupling of metal plates, namely horizontally [10] or vertically arranged [5]. The most promising ones are those vertically arranged, from both the space occupancy and performance points of view. Recent solutions are based on four-plate and six-plate vertically aligned architectures. The four-plate arrangement is described in the following.

Figure 3.2 shows the perspective view of the horizontally arranged plates.

Figure 3.3 reports a perspective view of a vertically aligned four-plate CPT. It consists of one transmitting plate, P_1, one receiving plate, P_2 and two shielding plates, S_1, S_2.

Figure 3.4 shows the cross section of a vertically arranged four-plate configuration with the main geometrical dimensions.

It is more convenient to arrange the plates vertically to save space. The two plates at the same side are of different sizes, and the shielding plate is larger to maintain the coupling with the plates at the other side. The length of P_1 and P_2 is l_1, the length of S_1 - S_2 is l_2, the distance between P_1 - P_2 is d, which is the air gap between the transmitting and receiving sides, and the distance S_1 - P_1 (S_1 - P_1) is d_{SP}. At each side, the distance d_{SP}

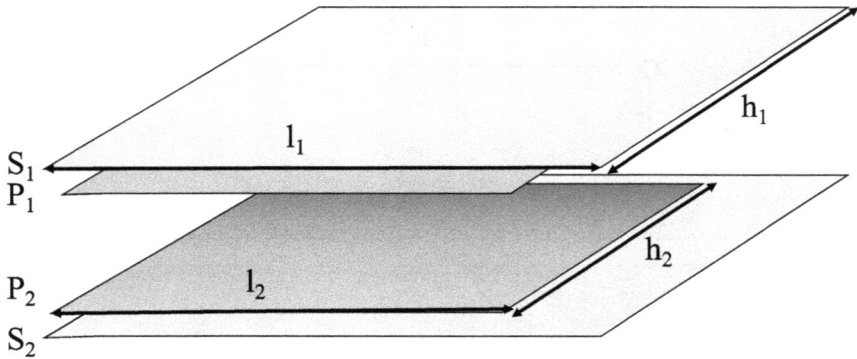

FIGURE 3.3   Perspective view of a vertically aligned four-plate CPT.

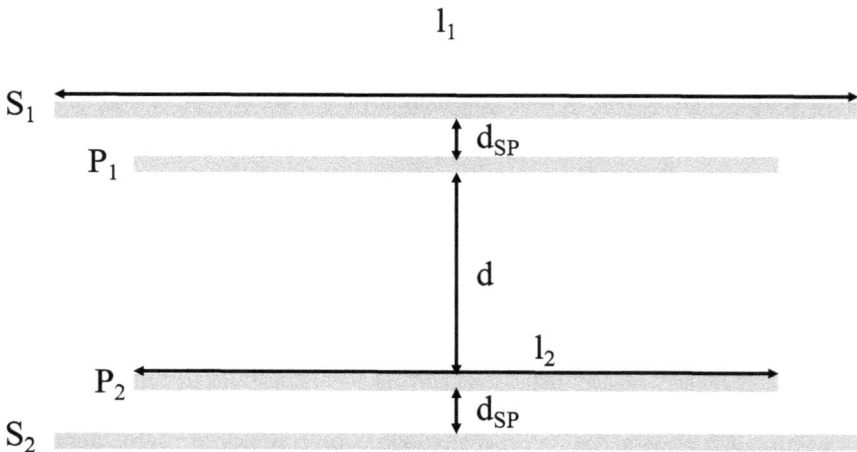

FIGURE 3.4   Cross-section view of a vertically aligned four-plate CPT.

must be kept low to maintain a large coupling capacitance and this can be regulated during the overall optimization of the CPT link. It has been demonstrated that the capacitance does not relate to the misalignment, so the system is robust with respect to misalignment [5]. The considered topology is fully symmetrical, for the transmitting and receiver side, but this constraint can be removed for the final design of the link, and different values for the geometrical parameters can be used for the fine-tuning of the entire system.

All the geometrical parameters must be optimized in such a way to determine the desired CPT link input/output relationships. For instance, while the plates P_1 and P_2 and the distance between them are mainly responsible for the input–output coupling, the shielding plates affect the electric field between the sending or receiving plate on the same side in such a way that the distances between the shielding plates and the sending or receiving plates can be used to fine-tune the input–output relationships.

Let us first consider the equivalent circuit model of the four-plate CPT link [5], assuming the absence of lossy contributions: it consists of a total of six coupling capacitors, since every two plates are capacitively coupled. The resulting network of equivalent capacitances is shown in Figure 3.5.

In order to rigorously compute the two-port network CPT link, two independent voltages are applied on the plates P_1, P_2, V_1, V_2, the corresponding currents, I_1, I_2, are injected into the plates from the transmitting and receiving sides, as shown in Figure 3.5.

By applying Kirchhoff's current law, the voltage and current relationships may be computed:

$$V_1 = I_1 \cfrac{1}{j\omega\left[C_{S2-P2} + \cfrac{\left(C_{S12}+C_{S2-P1}\right)\left(C_{S1-P2}+C_{P12}\right)}{C_{S12}+C_{S2-P1}+C_{S1-P2}+C_{P12}} \right]} +$$

$$V_2 \cfrac{C_{P12}C_{S12} - C_{S2-P1}C_{S1-P2}}{C_{S2-P2}\left(C_{S12}+C_{S2-P1}+C_{S1-P2}+C_{P12}\right)+\left(C_{S12}+C_{S2-P1}\right)\left(C_{S1-P2}+C_{P12}\right)}$$

$$(3.2)$$

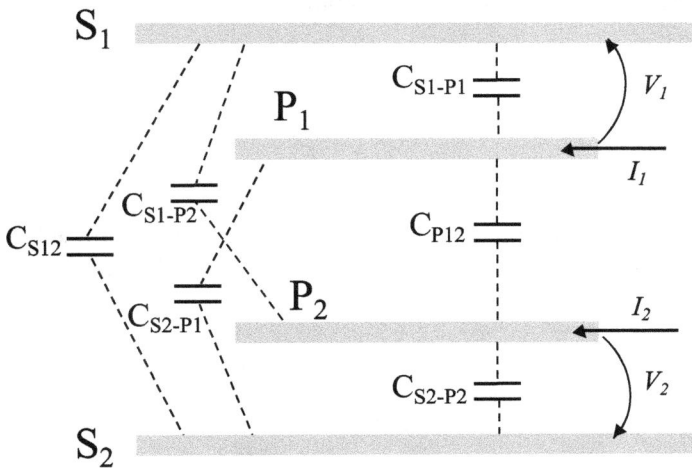

FIGURE 3.5   Circuit model of the capacitive coupling in a four-plate CPT link.

$$V_2 = I_2 \cfrac{1}{j\omega \left[C_{S1-P1} + \cfrac{\left(C_{S12} + C_{S1-P2}\right)\left(C_{S2-P1} + C_{P12}\right)}{C_{S12} + C_c + C_{S1-P2} + C_{P12}} \right] +}$$

$$V_1 \cfrac{C_{P12}C_{S12} - C_{S2-P1}C_{S1-P2}}{C_{S1-P1}\left(C_{S12} + C_{S2-P1} + C_{P12} + C_{S1-P2}\right) + \left(C_{S12} + C_{S1-P2}\right)\left(C_{S2-P1} + C_{P12}\right)}$$

$$(3.3)$$

From Equations (3.2, 3.3) the capacitances C_{TX}, C_{RX}, and C_M are defined as:

$$C_{TX} = C_{S1-P2} + \frac{\left(C_{S12} + C_{S2-P2}\right)\left(C_{S1-P2} + C_{P12}\right)}{C_{S12} + C_{S2-P2} + C_{S1-P2} + C_{P12}} \qquad (3.4)$$

$$C_{RX} = C_{S1-P1} + \frac{\left(C_{S12} + C_{S1-P2}\right)\left(C_{S2-P2} + C_{P12}\right)}{C_{S12} + C_{S2-P1} + C_{S1-P2} + C_{P12}} \qquad (3.5)$$

$$C_M = \frac{C_{24}C_{S12} - C_{S2-P2}C_{S1-P2}}{C_{S1-P1}\left(C_{S12} + C_{S2-P2} + C_{24} + C_{23}\right) + \left(C_{S12} + C_{S1-P2}\right)\left(C_{S2-P2} + C_{P12}\right)}$$

$$(3.6)$$

The coupling coefficient, which is a performance index highly relevant for predicting the CPT link capabilities, as for the IPT counterpart is defined as:

$$k_c = \frac{C_M}{\sqrt{C_{RX}C_{TX}}} \qquad (3.7)$$

Rearranging Equations (3.4), (3.5), and (3.6), a more compact representation of the voltages and currents relationships at the CPT link ports is obtained:

$$\begin{aligned} V_1 &= I_1 \cdot \frac{1}{j\omega C_{TX}} + V_2 \cdot \frac{C_M}{C_{TX}} \\ V_2 &= I_2 \cdot \frac{1}{j\omega C_{TX}} + V_1 \cdot \frac{C_M}{C_{RX}}. \end{aligned} \qquad (3.8)$$

and the Y-matrix representation of the CPT link can be derived from:

$$I_1 = j\omega C_{TX} \cdot V_1 - j\omega C_M \cdot V_2$$
$$I_2 = -j\omega C_M \cdot V_1 + j\omega C_{RX} \cdot V_2 \tag{3.9}$$

by setting:

$$C_1 = C_{TX} - C_M$$
$$C_2 = C_{RX} - C_M \tag{3.10}$$
$$C_3 = C_M$$

The resulting equivalent circuit of the lossless capacitive wireless link is reported in Figure 3.6(a). Since the shielding plates affect the electric field distribution between the sending or receiving plates on the same side, the distances between the shielding plates and the sending or receiving plates can be used to adjust the values of C_1 and C_2. To account for the losses of the plates and of the materials embedding the plates, the circuit can be modified as in Figure 3.6(b), by adding for each capacitance branch a proper conductance equivalent to the losses. Conductance values may be established by full-wave analysis of the CPT link including the realistic description of the electro-magnetic parameters of the metal plates and of the surrounding materials [5].

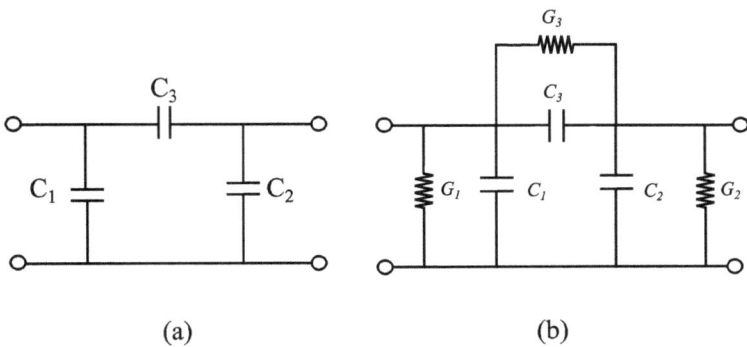

(a) (b)

FIGURE 3.6   Pi equivalent network representation of the capacitive CPT link (a) and of the capacitive link accounting for the metal plates and media losses (b).

3.2.3 Operating Conditions of the Wireless Capacitive Link

In practical applications, the system in Figure 3.1 is adopted for different operating conditions, depending on the applications, and a configuration that guarantees maximum efficiency may be preferred to an alternative, ensuring maximum power absorbed by the load at the expense of lower efficiency, and vice versa. In the following, a rigorous theoretical method to the design alternatives for the RF CPT sub-system of Figure 3.1 is presented [11].

Through circuit theory [11], a general-purpose approach for the simultaneous determination of its RF terminations, namely the RF source excitation at the transmitting side and the loading subnetwork at the receiving side, is proposed.

Figure 3.7 reports the equivalent Pi-network of the RF CPT subsystem of Figure 3.6(b), embedded between its complex transmitting and receiving terminations. A current source with internal complex admittance $Y_{c1} = G_{c1} + jB_{c1}$, represents the Norton equivalent of the connection between the DC-RF converter and the TX matching/compensating network; the complex admittance $Y_{c2} = G_{c2} + jB_{c2}$, loading the capacitive link,

RF CPT sub-system

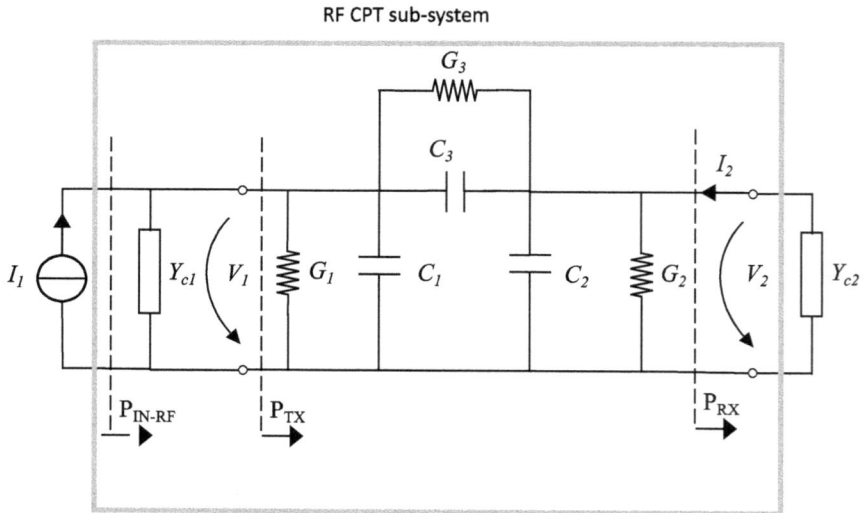

FIGURE 3.7 Equivalent network representation of the capacitive CPT link in Figure 3.6: the power source and the compensating network are represented by the current source I_1 with internal complex admittance $Y_{c1} = G_{c1} + jB_{c1}$; the receiver compensating network loaded by the RF-DC converter is represented by the complex load $Y_{c2} = G_{c2} + jB_{c2}$.

is the circuit equivalent of the connection between the RF matching/compensating network and the loaded RF-DC converter.

Once the metal plates architecture has been defined and its P1-equivalent network representation has been computed, the problem to be solved is finding the best values of the real and imaginary parts of $Y_{cj} = G_{cj} + jB_{cj}$, $j = 1,2$, for realizing the capacitive wireless power transfer at various operating conditions. In particular it will be shown that different values are needed, $Y_{cj} = G_{cj} + jB_{cj}$, $j = 1,2$, depending on the following final goals:

A. to obtain the maximum efficiency, η_{RF-RF};
B. to obtain the maximum power, P_{OUT-RF};
C. to obtain the conjugate matching conditions (i.e. power matching).

3.2.4 Computation of the Power Loss Contributions Inside the WPT Capacitive Network

Let us consider the reciprocal two-port network of Figure 3.6(b) which is the equivalent of the WPT capacitive wireless link represented by its admittance matrix:

$$Y = \begin{bmatrix} y_{11} & y_{12} \\ y_{12} & y_{21} \end{bmatrix} \tag{3.11}$$

where $y_{ij} = G_{ij} + jB_{ij}$ ($i,j = 1,2$). It consists of the Pi equivalent network (Figure 3.6(b)) and the admittance matrix is:

$$Y = \begin{bmatrix} Y_1 + Y_3 & -Y_3 \\ -Y_3 & Y_2 + Y_3 \end{bmatrix} \tag{3.12}$$

With the admittance terms in Equation (3.11) given by:

$$y_{11} = (G_1 + G_3) + j\omega(C_1 + C_3)$$
$$y_{22} = (G_2 + G_3) + j\omega(C_2 + C_3) \tag{3.13}$$
$$y_{12} = -G_3 - j\omega C_3.$$

While the metal plates losses are accounted for, the dielectric losses can be negligible and $G_3 = 0$ can be assumed.

From Figure 3.6(b), V_1 and V_2 can be computed as:

$$V_1 = \frac{y_{22} + Y_{c2}}{\Delta'} I_1 \tag{3.14}$$

and

$$V_2 = -\frac{y_{12}}{y_{22} + Y_{c2}} V_1 \tag{3.15}$$

where:

$$\Delta' = (y_{11} + Y_{c1})(y_{22} + Y_{c2}) - y_{12}^2 \tag{3.16}$$

Using these quantities, the power contributions involved can be computed. The available RF power P_{AV} from the $DC-RF$ converter:

$$P_{AV} = \frac{I_1^2}{8G_1} \tag{3.17}$$

is absorbed by the system through the following three mechanisms:

Power absorbed by G_1:

$$P_{G_1} = \frac{1}{2}(G_1 + G_{c1})|V_1|^2 = \frac{1}{2}(g_1 + G_{c1})\frac{|y_{22} + Y_{c2}|^2}{|\Delta'|^2}|I_1|^2 \tag{3.18}$$

Power absorbed by G_2:

$$P_{G_2} = \frac{1}{2}\frac{g_{22}|y_{12}|^2|I_1|^2}{|\Delta'|^2} \tag{3.19}$$

Power absorbed by the load G_{c2}:

$$P_{G_{c2}} = \frac{1}{2} \frac{G_{c2} |y_{12}|^2 |I_1|^2}{|\Delta'|^2} \qquad (3.20)$$

Using these quantities the efficiency is expressed by the following ratio:

$$\eta_{RF-RF} = \frac{P_{G_{c2}}}{P_{G_1} + P_{G_2} + P_{G_{c2}}} \qquad (3.21)$$

which, by substituting Equations (3.18)–(3.20) into Equation (3.21) becomes:

$$\eta_{RF-RF} = \frac{G_{c2} G_1 G_2 \chi^2}{|y_{22} + Y_{c2}|^2 (G_1 + G_{c1}) + G_1 G_2^2 \chi^2 + G_{c2} G_1 G_2 \chi^2} \qquad (3.22)$$

with the parameter χ, defined as:

$$\chi^2 = \frac{\omega^2 C_3^2}{G_1 G_2} \qquad (3.23)$$

It is noteworthy that χ may also be used as a figure of merit of the wireless capacitive link since it is strictly related to the coupling coefficient (3.7).

3.2.4.1 Maximization of the Efficiency: η_{RF-RF}

In this case Y_{c1} and Y_{c2} are found by equating the derivative of Equation (3.22) to zero.

The equivalent admittances become:

$$Y_{c1} = -j\omega(C_1 + C_3) \qquad (3.24)$$

$$Y_{c2} = G_2 \sqrt{1 + \chi^2} - j\omega(C_2 + C_3) \qquad (3.25)$$

and the expression for the maximum efficiency becomes:

$$\eta_{RF-RF} = \frac{\chi^2}{\left(1+\sqrt{1+\chi^2}\right)^2} \qquad (3.26)$$

which asymptotically, for $\chi \to \infty$, provides an efficiency value of $\eta_{RF-RF\infty} \to 1$.

In such conditions, $P_{G_{c2}}$, the power absorbed by the load is:

$$P_{G_{c2}} = 4P_{AV}\frac{\chi^2}{\sqrt{1+\chi^2}\left(1+\sqrt{1+\chi^2}\right)^2} \qquad (3.27)$$

which, asymptotically, for $\chi \to \infty$, gives a behavior of the type $P_{RF-RF'\infty} \to 4P_{AV}/\chi$.

3.2.4.2 Maximization of the Power Absorbed by the Load: $P_{G_{c2}}$

In this case, the expressions of Y_{cj} are found by equating the derivatives of Equation (3.20) to zero.

The equivalent admittances become:

$$Y_{c1} = G_{c1} + jB_{c1} = -j\omega(C_1 + C_3) \qquad (3.28)$$

$$Y_{c2} = G_{c2} + jB_{c2} = G_2(1+\chi^2) - j\omega(C_2 + C_3) \qquad (3.29)$$

It is noteworthy that only G_{c2} is changed with respect to the solution that maximizes efficiency. In such conditions, the efficiency becomes

$$\eta_{RF-RF} = \frac{\chi^2}{2(2+\chi^2)} \qquad (3.30)$$

which asymptotically, for $\chi \to \infty$, provides an efficiency value of $\eta_{2\infty} \to 1/2$.

The maximum of $P_{G_{c2}}$, the power absorbed by the load, is:

$$P_{G_{c2}} = P_{AV}\frac{\chi^2}{(1+\chi^2)} \qquad (3.31)$$

3.2.4.3 Conjugate Matching Solution

To compute the conjugate image admittances the procedure presented in Ref. [12] can be adopted. In this case the termination admittances take the form:

$$Y_{c1} = G_1\sqrt{1+\chi^2} - j\omega(C_1 + C_3) \tag{3.32}$$

$$Y_{c2} = G_2\sqrt{1+\chi^2} - j\omega(C_2 + C_3) \tag{3.33}$$

This network synthesis is very similar to the one resulting from the maximization of the efficiency, except for the value of G_{c1} which, in the present case, is different from zero.

In such conditions, the corresponding the efficiency is:

$$\eta_{RF-RF} = \frac{\chi^2}{2\left(1+\sqrt{1+\chi^2}\right)^2} \tag{3.34}$$

which, asymptotically, for $\chi \to \infty$, provides an efficiency value of $\eta_{2\infty} \to 1/2$.

The power absorbed by the load is:

$$P_{G_{c2}} = 4P_{AV}\frac{\chi^2}{\sqrt{1+\chi^2}\left(2+\sqrt{1+\chi^2}\right)^2} \tag{3.35}$$

which, asymptotically, for $\chi \to \infty$, gives a behavior of the type $P_{3\infty} \to 4P_0/\chi$.

Table 3.1 compares the synthesized admittances to be realized by the compensating networks at the transmitting and receiving sides, depending on the various specifications discussed above.

The real parts of the input and output terminations are strongly dependent on Equation (3.25) which affects the power absorbed through the entire CPT link, as has been derived in this section.

Note that when the wireless CPT link is excited by the actual power source at port 1, such as, e.g., a class E amplifier, G_{c1} is fixed by its nonlinear design.

TABLE 3.1   Relationship between the CPT Wireless Link Equivalent Circuit Parameters and the Terminations for Different Design Specifications

Specification	G_{c1}	B_{c1}	G_{c2}	B_{c2}
Efficiency	0	$-\omega(C_1+C_3)$	$G_2\sqrt{1+\chi^2}$	$-\omega(C_2+C_3)$
Power absorbed by the load	0	$-\omega(C_1+C_3)$	$G_2(1+\chi^2)$	$-\omega(C_2+C_3)$
Conjugate matching	$G_1\sqrt{1+\chi^2}$	$-\omega(C_1+C_3)$	$G_2\sqrt{1+\chi^2}$	$-\omega(C_2+C_3)$

The first and third rows, belonging to the highest efficiency and to the conjugate matching conditions, respectively, coincide except for G_{c1}. It also can be observed that the maximum efficiency solution results in a power on the load that reduces to zero as the quality of the link increases (i.e., for $\chi\to\infty$). This result is related to the definition assumed for the efficiency of the link. In fact, in order to maximize the efficiency of the link it is not necessary to maximize the absolute value of the power absorbed by the load but to maximize the portion of the power provided by the generator that is delivered to the load. Accordingly, the maximum efficiency solution can result in low values of the active power absorbed by the load.

3.3 NUMERICAL MODELING OF A CAPACITIVE WPT LINK

This section considers the practical implementation of a capacitive WPT link, its full wave simulation, and the computation of the equivalent circuit model. Results obtained by circuital and full-wave simulations are reported and compared with analytical data.

3.3.1 Full-wave Simulation of the Capacitive Link

A four-plate, vertically aligned, and fully symmetrical CPT link is shown in Figure 3.8 [11]. With reference to the schematic representation of Figures 3.3 and 3.4, the inner plates have dimensions (mm) $l_2 = 40$ and $h_2 = 25$, while the shielding plates have dimensions (mm) $l_1 = 50$ and $h_1 = 30$; the distance between the shielding and the active plates is $d_{sp} = 1.6$ and the plates distance is $d = 3.2$. The plate material adopted is copper and the medium embedding the link is dry air ($G_3 = 0$). The link is characterized to operate at a high frequency value of 15 MHz.

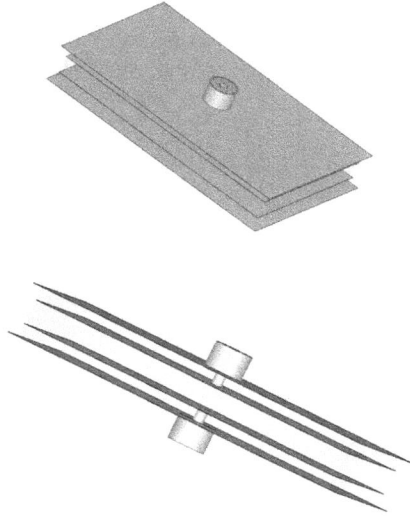

FIGURE 3.8   Perspective view (a) and cross view (b) of the simulated structure.

The CPT link ports are obtained by two 50 Ω coaxial cables with the inner conductor soldered to the inner plate P_1 (P_2) and the external conductor soldered to the shielding plates S_1 (S_2).

From the full-wave simulation results, the parameters of the equivalent pi-network of Figure 3.6(b) have been computed and are reported in Table 3.2. For this set up, $\chi = 198.8$ at 15 MHz. This table also reports the computed values of the transmitting and receiving complex admittance terminations derived for the cases discussed in subsections: 3.2.4.1, 3.2.4.2, and 3.2.4.3.

3.3.2 Circuital Simulation of the Capacitive Link

By considering an equivalent current generator of 1 mA, as the source excitation of the wireless CPT link, the corresponding values of the efficiency and power absorbed by the load (both analytically computed and derived from circuit simulation) are reported in Table 3.3, for the three possible design goals discussed in the above-cited subsections.

The same circuit simulation allows to compute the CPT link performance with respect to any possible load: $R_L = \dfrac{1}{G_{C2}}$. The results are plotted in

TABLE 3.2 Equivalent Circuit Parameters of the Wireless CPT Link Derived from Full-wave Simulation: Susceptibilities Coincide for any Design Case and thus the Resulting Inductances. Different Loading Conditions are Predicted for the Cases A and B Discussed in Section 3.2.3

Parameter	Value
$C_1 = C_2$	6.771 pF
C_3	0.784 pF
$G_1 = G_2$	5.242024 μS

Parameter	Computed value
$B_{C1} = B_{C2}$	-713 μS
$(L_1 = L_2)$	14.89 μH
R_{L_A}	13,499.65 Ω
R_{L_B}	955.31 Ω

TABLE 3.3 Analytical and Circuital Simulation Results Obtained for the Structure Illustrated in Figure 3.8. For the Three Analyzed Solutions a Perfect Agreement has been Obtained between Data Calculated by using the Analytical Formulas and Circuital Simulation Results

Parameter	Analytical results	Circuital simulations
η_A	0.868	0.868
P_A (mW)	5.858	5.858
η_B	0.495	0.495
P_B (mW)	23.644	23.644
η_C	0.434	0.434
P_C (mW)	5.136	5.136

Figure 3.9, they show that the efficiency is maximized for $R_L = 13499.65\Omega$ while the power absorbed by load is maximized for $R_L = 955.31\Omega$, thus confirming the analytical data given in Table 3.2.

Further simulations have been performed in order to evaluate the frequency behavior of the link; the obtained results are reported and discussed in the following part of this section.

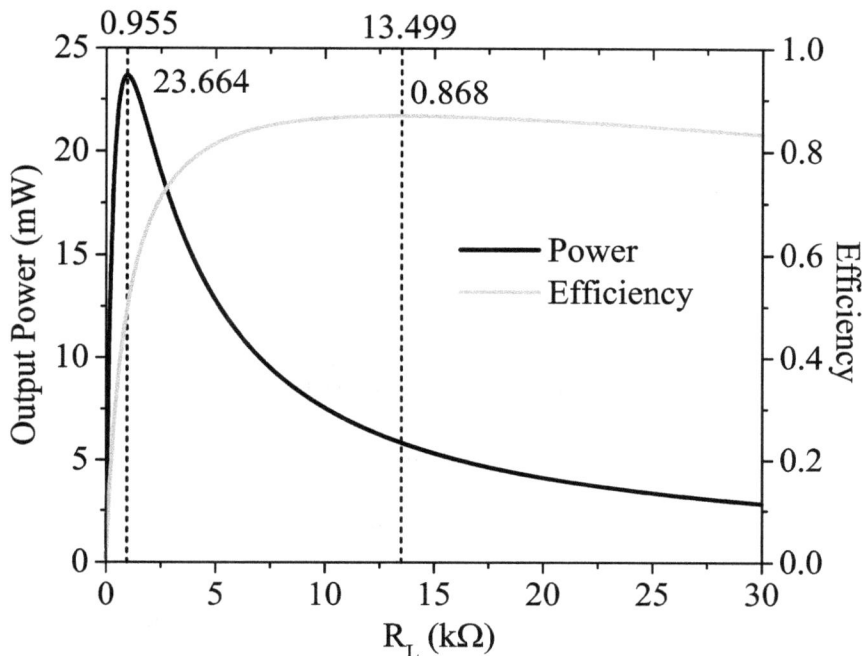

FIGURE 3.9   Predicted output power and efficiency as a function of the load impedance.

The same circuit simulation allows to predict the behavior of the CPT link, terminated by the designed Y_{c1} and Y_{c2} with respect to the operating frequency. This analysis is very relevant to determine not only the link operating band, but also its tuning capabilities.

In particular, the two network solutions satisfying the design goals: (A) to obtain the maximum efficiency η_{RF-RF} or (B) to obtain the maximum power P_{OUT-RF}, are considered. Figures 3.10 and 3.11 show the frequency behavior of the efficiency (η_{RF-RF}) and of the power absorbed by the load (P_{OUT-RF}), respectively. As expected, the network corresponding to goal B, shows a 50% efficiency at the operating frequency of 15 MHz, and reaches saturation for increasing frequency; while the network for goal A shows a peak at the operating frequency.

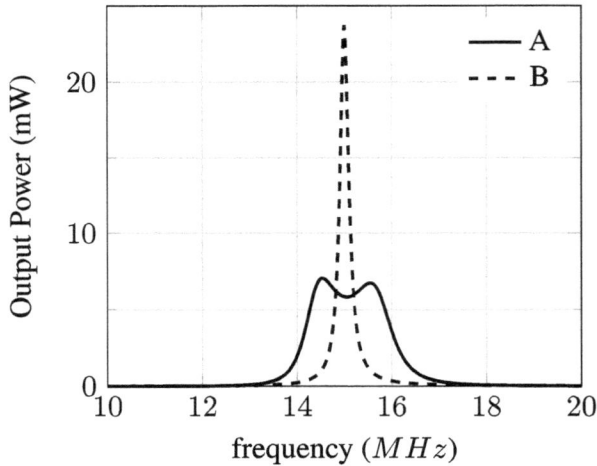

FIGURE 3.10 P_{OUT-RF} obtained by circuital design A to obtain the maximum efficiency η_{RF-RF}, and B to obtain the maximum power P_{OUT-RF}.

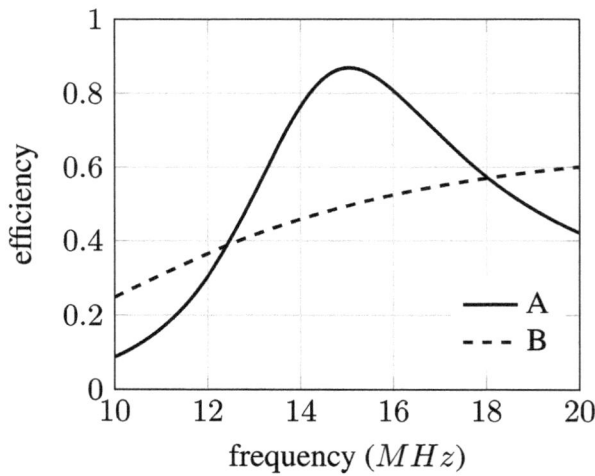

FIGURE 3.11 η_{RF-RF} obtained by circuital design: A to obtain the maximum efficiency η_{RF-RF}, and B to obtain the maximum power P_{OUT-RF}.

Figures 3.12–3.14 show the corresponding trends of the input and output voltages, V_1 and V_2, and of the output current I_2. It is noteworthy that when goal A is concerned a much higher output voltage V_2 (lower output current I_2) is required with respect to goal B, and vice versa.

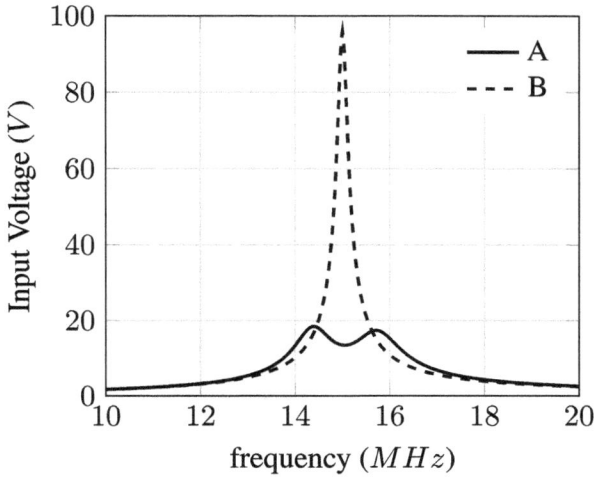

FIGURE 3.12 V_1 obtained by circuital design: A to obtain the maximum efficiency η_{RF-RF}, and B to obtain the maximum power P_{OUT-RF}.

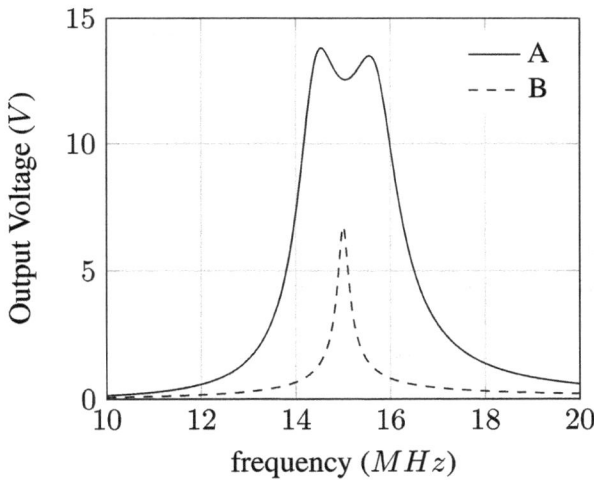

FIGURE 3.13 V_2 obtained by circuital design: A to obtain the maximum efficiency η_{RF-RF}, and B to obtain the maximum power P_{OUT-RF}.

3.4 COMPENSATING NETWORKS

In practical implementations of CPT links, the terminations Y_{C1} and Y_{C2}, whose design specifications have been discussed in the previous section, must be implemented by a real sub-system on the transmitting and receiving sides.

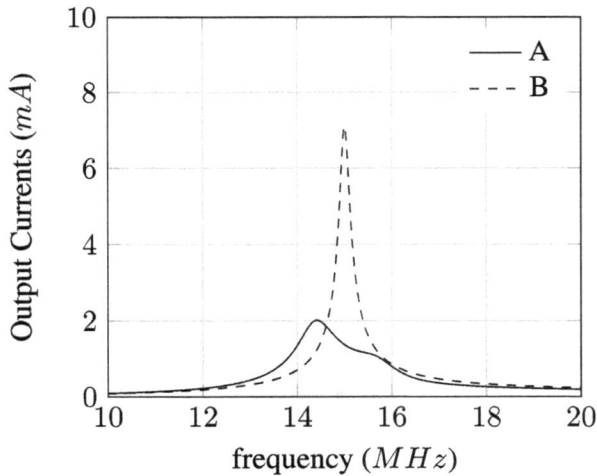

FIGURE 3.14 I_2 obtained by circuital design: A to obtain the maximum efficiency η_{RF-RF}, and B to obtain the maximum power P_{OUT-RF} .

On the transmitter side, they are realized by the connection of an inverter with a compensating network of the wireless CPT link.

Similarly, on the receiver side, they consist of a compensating network connected to a rectifier. As stated in the previous sections, the main challenge is to define these sub-systems accounting for the reciprocal influences, being the entire CPT system coupled.

As for IPT systems, these subnetworks may be realized by non-resonant and resonant topologies. The non-resonant topology consists of an inverter, such as a converter, and the coupling capacitance link works as a power storage components to smooth the power in the circuit. In such conditions, large capacitances are needed, usually in the 10s of nanofarad range, and the transferred distance must be kept very low (of the order of mm).

The resonant topologies consist of inverters with inductive terminations which resonate with the coupling capacitors. In this case the transmitting side must be designed accounting for the accurate description of the resonating system. The same is true for the receiving side. In this operating conditions, there is more flexibility concerning the design of the wireless CPT, which needs to operate with lower capacitance values but the resonating inductance values increase, thus increasing the losses contributions (G_1 and G_2), and the operating frequency should be increased.

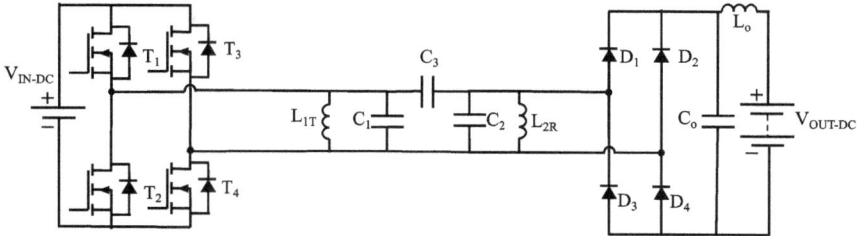

FIGURE 3.15 A simple implementation of the end-to-end CPT link of Figure 3.1.

Furthermore, if resonant topologies are chosen, their performances are more dependent on the CPT link set-up and the end-to-end performance may be affected by slight variations of the link operating conditions, such as misalignment or metal plates distance variations or changing of the materials characteristics located between them. The simplest compensated network is shown Figure 3.15. A full-bridge inverter is used as the primary side power source, while a full-bridge rectifier is adopted as the secondary side power conditioning circuit, and a low-pass filter connects the rectifier output with the . In this chapter, only a half-bridge is used to reduce the system cost. Full-bridge topologies are more suitable for high-power CPT links (of the order of kW) while half-bridge ones are used for lower power transfer. Both at the receiver and transmitter sides, two parallel connected inductances are used to resonate the CPT link. Several other arrangements can be conceived of this simple compensating network.

More complex solutions have been tested for high-power transfer, such as a double-sided LCLC-compensated circuit for high-power and large air gap applications. A vertical structure of plates and an LCL compensation circuit is used: 1.87-kW output power and 85.87% efficiency with 150-mm air gap has been demonstrated. An interesting topology has been studied for railway applications [13]. An interesting topology has been studied for railway applications to eliminate the pantograph and contact wire. In this case, the chassis of the vehicle is connected to the track and ground. The track and ground are used as a circuit return path, and are used as the shielding plates of the CPT link. Thus a CPT system for a railway vehicle only requires two plates. A MW power transfer at 200 kHz is predicted and demonstrated with a scaled prototype.

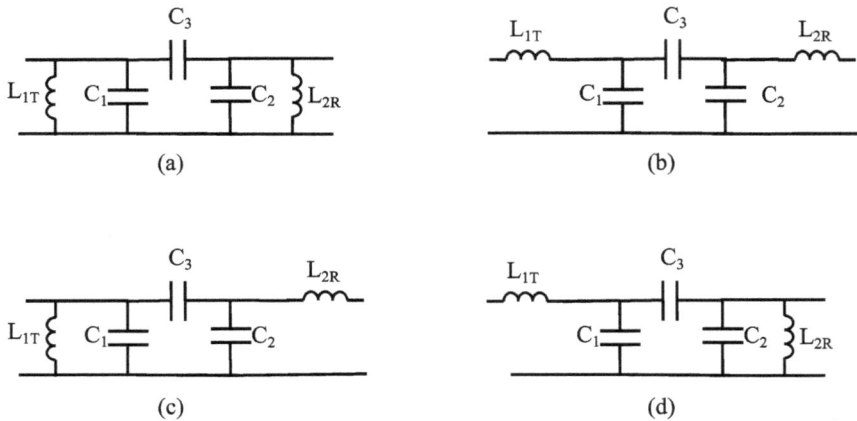

FIGURE 3.16 Alternatives for the compensating networks. (a) parallel-parallel; (b)series-series; (c) parallel-series; (d) series-parallel.

3.5 CONCLUSIONS

This chapter has been dedicated to the analysis and design of a CPT link. First the link has been analyzed from the circuital point of view, and a two-port network equivalent of the link has been analytically derived. Then, the link has been embedded inside source and load equivalent circuits and the general-purpose procedure to design the entire system has been developed showing that different goals and operating condition needs may be carefully accounted for. It has been demonstrated analytically that the operating conditions for a CPT end-to-end system significantly differ when the main goal is efficiency or power absorbed by the load. The procedure has been developed accounting for the possible loss contributions that may be present. It has been found that the maximum efficiency solution coincides with conjugate matching when it is possible to realize on the transmitter port using a zero input resistance on the transmitter port. In addition, it has been shown that the maximum efficiency solution provides decreasing values of power as the efficiency is increased. On the contrary, the maximum power approach realizes a maximum efficiency of 50% but does not limit the amount of power deliverable to the load. Some recent examples taken from the literature for the effective realization of the compensating networks have been proposed and described.

REFERENCES

[1] N. Tesla, "Experiments with Alternate Currents of Very High Frequency and Their Application to Methods of Artificial Illumination," *Transactions of the American Institute of Electrical Engineers*, vol. VIII, no. 1, pp. 266–319, 1891.

[2] T. C. Martin, *The Inventions, Researches and Writings of Nikola Tesla*, 1977. Barnes & Noble Books (1995).

[3] H. Zheng, K. Tnay, N. Alami, and A. Hu, "Contactless Power Couplers for Respiratory Devices," *Proceedings of 2010 IEEE/ASME International Conference on Mechatronic and Embedded Systems and Applications*, QingDao, China, pp. 155–160, 2010, doi: 10.1109/MESA.2010.5552082.

[4] A. M. Sodagar and P. Amiri, "Capacitive Coupling for Power and Data Telemetry to Implantable Biomedical Microsystems," *4th International IEEE/EMBS Conference on Neural Engineering*, Antalya, Turkey, pp. 411–414, 2009, doi: 10.1109/NER.2009.5109320.

[5] H. Zhang, F. Lu, H. Hofmann, W. Liu, and C. C. Mi, "A Four-Plate Compact Capacitive Coupler Design and Lcl-Compensated Topology for Capacitive Power Transfer in Electric Vehicle Charging Application," *IEEE Transactions on Power Electronics*, vol. 31, no. 12, pp. 8541–8551, 2016.

[6] Chieh-Kai Chang, G. G. Da Silva, A. Kumar, S. Pervaiz, and K. K. Afridi, "30 W Capacitive Wireless Power Transfer System with 5.8 pF Coupling Capacitance," *2015 IEEE Wireless Power Transfer Conference (WPTC)*, Boulder, CO, USA, pp. 1–4, 2015, doi: 10.1109/WPT.2015.7140184.

[7] C.-K. Chang, G. G. Da Silva, A. Kumar, S. Pervaiz, and K. K. Afridi, "30 w Capacitive Wireless Power Transfer System with 5.8 pf Coupling Capacitance," in *2015 IEEE Wireless Power Transfer Conference (WPTC)*, pp. 1–4, 2015.

[8] J. Estrada, S. Sinha, B. Regensburger, K. Afridi, and Z. Popović, "Capacitive Wireless Powering for Electric Vehicles with Near-field Phased Arrays," *2017 47th European Microwave Conference (EuMC)*, Nuremberg, Germany, pp. 196–199, 2017, doi: 10.23919/EuMC.2017.8230833.

[9] A. Costanzo, M. Dionigi, D. Masotti, M. Mongiardo, G. Monti, L. Tarricone, and R. Sorrentino, "Electromagnetic Energy Harvesting and Wireless Power Transmission: A Unified Approach," *Proceedings of the IEEE*, vol. 102, no. 11, pp. 1692–1711, 2014.

[10] F. Lu, H. Zhang, H. Hofmann, and C. Mi, "A Double-Sided Lclc-Compensated Capacitive Power Transfer System for Electric Vehicle Charging," *IEEE Transactions on Power Electronics*, vol. 30, no. 11, pp. 6011–6014, 2015.

[11] M. Dionigi, M. Mongiardo, G. Monti, and R. Perfetti, "Modelling of Wireless Power Transfer Links Based on Capacitive Coupling: Rigorous Analysis of Wpt Systems Based on Capacitive Coupling," *International Journal of Numerical Modelling: Electronic Networks, Devices and Fields*, vol. 30, no. 8, e2187, 2016.

[12] S. Roberts, "Conjugate-Image Impedances," in *Proceedings of the IRE*, vol. 34, no. 4, pp. 198p–204p, April 1946, doi: 10.1109/JRPROC.1946.234242.

[13] S. Li, Z. Liu, H. Zhao, L. Zhu, C. Shuai, and Z. Chen, "Wireless Power Transfer by Electric Field Resonance and Its Application in Dynamic Charging," *IEEE Transactions on Industrial Electronics*, vol. 63, no. 10, pp. 6602–6612, 2016.

CHAPTER 4

The Receiving Part of Radiative Wireless Power Transfer

Nuno Borges Carvalho

4.1 INTRODUCTION

Wireless power transfer (WPT) has been researched since Heinrich Hertz first experimented with radio waves in the 19th century. However, achieving high reception efficiencies is challenging when designing a WPT system. Moreover, the transmitted beam should be focused on improving the efficiency of the transmitted electromagnetic wave.

While working for Raytheon Company, William C. Brown demonstrated the possibility of designing highly efficient rectennas in the 1970s. By using a diode that combined Pt/GaAs, Brown achieved efficiencies of over 90% [1].

Nevertheless, the inherent difficulty in achieving highly efficient WPT systems has impeded their generalized application in real-life situations. Although problems in the generation and reception of microwaves have been minimized, the propagating beam divergence remains a considerable problem. At the University of Aveiro, research has been pursued into all the WPT system components: DC-RF conversion, antennas and beam propagation, and RF-DC conversion.

This chapter focuses on the work done at the receiving part, especially the RF-DC conversion and its application in high power and mm-wave

DOI: 10.1201/9781003328636-4

frequencies. These results can help the development of new solutions for space-based solar power satellites and other applications.

4.2 RF-DC CONVERSION

Nowadays, all electric devices store their power in a battery. With the increase in the demand for Internet of Things (IoT) equipment, two problems arise: the need for bigger batteries [2] and the amount of power consumed [3]. According to Ref. [4], in 2025, it is estimated that more than 12 billion batteries per year will be used. This brings an environmental sustainability problem because the resources to build batteries with ample storage capacity are limited and not renewable. They take many years to decompose, at least with this day's knowledge, leading to severe environmental hazards [4]. There are already many solutions to this, one of which is reducing the battery size. Still, more solutions are needed. Therefore, to provide more liability, the batteries need to be charged faster and with more periodicity; for this, the best solutions are wireless power transmissions (WPT) and energy harvesting (EH) methods. Even though this is not the fastest charging method, it allows the equipment to be set up while moving. Moreover, with bigger batteries, the equipment would also have to be bigger, so by using WPT technologies, there is the potential for equipment to become smaller and more portable.

In recent years, some works related to RF-DC at mm-wave frequencies have been reported, for example, some of those that follow the points mentioned above were selected below. In Ref. [21], a fully integrated GaAs single-diode RF-DC was developed for the frequency of 5.8 GHz, achieving a maximum of 67% efficiency for 30 dBm of input power. In Ref. [22], a self-biased rectifier was designed and achieved 42% efficiency with 12.5 dBm of input power. The authors of Ref. [23] show a GaAs rectenna following a class-F amplifier operation that resulted in good efficiency only for high input power values. An inkjet-printed rectenna was presented in Ref. [24], resulting in a new and non-conventional approach to developing this system. However, the efficiency performance was below 40% in the analyzed sweep. Finally, the theoretical diode behavior background is explained in Ref. [25] for a 35-GHz rectenna. Its experimental prototype showed an efficiency of 60% for 21 dBm of input power. All these works present good efficiencies at high input power, which can sometimes be very difficult to implement in a low-power scenario.

There are two main divisions in the WPT methods, the near field, which can achieve very high efficiencies, but with the downside of its efficiency decreasing significantly with the distance between transmitter and receiver, and the far field, which has lower efficiency values but has the advantage of allowing the devices to be truly wireless, which is why the devices with far-field technology in the market are increasing significantly and are provisioned to grow exponentially over the next few years [5], being found in the most common appliances, such as cars, mobile phones, etc. The WPT system can be divided into three parts (Figure 4.1): the transmitter, the microwave beam, and the receiver. The DC-DC efficiency of the overall system depends on the efficiency of each of those blocks, and even the slightest diversion of this value will significantly impact the final efficiency. Therefore, to prevent high drops in efficiency, it is essential to improve each section of the system. The EH method is based on WPT technology, but instead of using a transmitter to supply the receiver with power, it harvests the electromagnetic waves available. This is only possible due to the increased radio frequency (RF) microwaves surrounding us [6].

In both WPT and EH, the receiver block can be composed of a simple rectenna, a concept proposed in 1960 by William C. Brown [7]. As this is not a new concept, there already exists a vast set of articles discussing this theme and providing improvements to the rectenna's circuit. One of the most important parts when designing any circuit, including a rectenna, is correctly matching the impedance. When it comes to the rectenna's circuit, the diode is the component that makes impedance matching very important, as its behavior improves significantly when the correct impedance is used. To match the impedance, it is common to use the simulator to find the impedance that provides the highest efficiency to the converter or to use the complex conjugate to match the impedance in both sides of the diode.

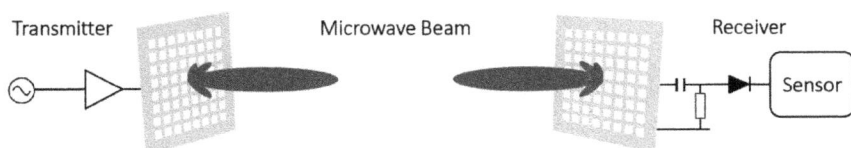

FIGURE 4.1   WPT's block diagram.

4.2.1 High Input Impedance Efficiency Enhancement of RF-DC Converter

Even though the traditional method to design the input impedance block in a radio frequency to direct current (RF-DC) converter is to match the diode's source and load impedance through the complex conjugate, it is possible to analyze it in some papers, such as Refs. [11–13]. The most common method found in newer articles is to use a source pull simulation to find the best impedance for the diode used in each RF-DC converter or rectenna. Through the analysis of some papers, it is possible to conclude that depending on the configuration of the diode, the source impedance should have different values. First, if a bridge or shunt configuration is used the input impedance should be as high as possible, as it is possible to see in some articles, such as [9,10,15–17,19], in which the input impedance seen from the diode is always higher than 100 Ω. Second, if a voltage doubler configuration is used, then the input reactance should be high, as it is possible to observe in Refs. [14,18]. A brief overview of the papers cited in this paragraph can be found in Table 4.1. Regardless of these papers using a source pull method to design the input impedance block, there is still the need for an explanation as to why the diode requires a high impedance or reactance from the source. Therefore, a study is presented on the impacts of the source's impedance on the diode's performance when using a single shunt configuration.

With the intent of understanding how the impedance seen by the diode impacts its behavior and which impedance would be best to improve the diode, this work provides a theoretical and experimental study on the diode's

TABLE 4.1 Brief Overview of the Input Impedance Used in RF-DC Converters and Rectennas when a Source Pull Method Is Applied

Rectifier's configuration	Frequency	Impedance	Efficiency	References
Single shunt	915 MHz	0.94/9°	40.7%	[9]
Single shunt	2.45 GHz	0.94/24°	56.2%	[9]
Single shunt	900 MHz	181 + j375 Ω	70.6%	[10]
Single shunt	1.85 GHz	325 – j57 Ω	65%	[15]
Single shunt	1.85 GHz	349 – j166 Ω	70%	[15]
Single shunt	2.4 GHz	80 Ω	75%	[17]
Bridge	5.8 GHz	580 Ω	90%	[16]
Bridge	2.4 GHz	670 Ω	81%	[19]
Voltage doubler	930 MHz	20 + j280 Ω	70%	[14]
Voltage doubler	860 MHz	25 + j340 Ω	83%	[18]

input impedance to understand which input impedance improves the diode's behavior the most. For this, the diode will be tested in a load/source pull simulation, in which the impedance on both of the diode's ports will be changed to find the impedance which improves the converter's efficiency. To verify the simulation results, two circuits were designed and compared, one using the complex conjugate to match the impedance in both ports of the diode, the other using the impedance from the source pull simulation.

4.2.1.1 RF-DC Converters Design Theory

As explained previously, the lack of available power to charge wireless devices leads to a high demand for improved WPT systems. Thus, to achieve this goal, it is important to consider that the total efficiency of a WPT system is given by Equation (4.1). This means that a slight change in efficiency in one of the WPT sections can have a considerable impact on the overall system performance. Therefore, to prevent this from happening, improvement of all WPT sections is highly important.

$$\eta_{total} = \frac{P_{DCout}}{P_{DCin}} = \eta_{transmitter} * \eta_{Beam} * \eta_{receiver} \tag{4.1}$$

The next subsection focuses on the receiver part where the theory of both circuit and device design level will be presented.

4.2.1.1.1 Circuit-level Design Theory As its name indicates, a rectenna is an antenna with rectifying capabilities. It can be divided into two blocks, the antenna and the RF-DC converter. The latter's performance depends mainly on the rectifier's behavior. In order to rectify the signal, the most used component is the diode due to its very small number of parasitic elements, meaning it has a very low loss when compared to other components. Nevertheless, the diode has some disadvantages, the main one being the interconnection of its two ports, meaning that what happens in one side of the diode will interfere and affect the behavior on the other side. Therefore, if the antenna is directly connected to the diode, there is the need to create the best antenna to improve the diode's performance.

According to Equation (4.2), the characteristic curve of the diode is exponential, thus, when there is a small change in the diode's input signal, the output voltage is going to vary significantly, as demonstrated in Figure 4.2.

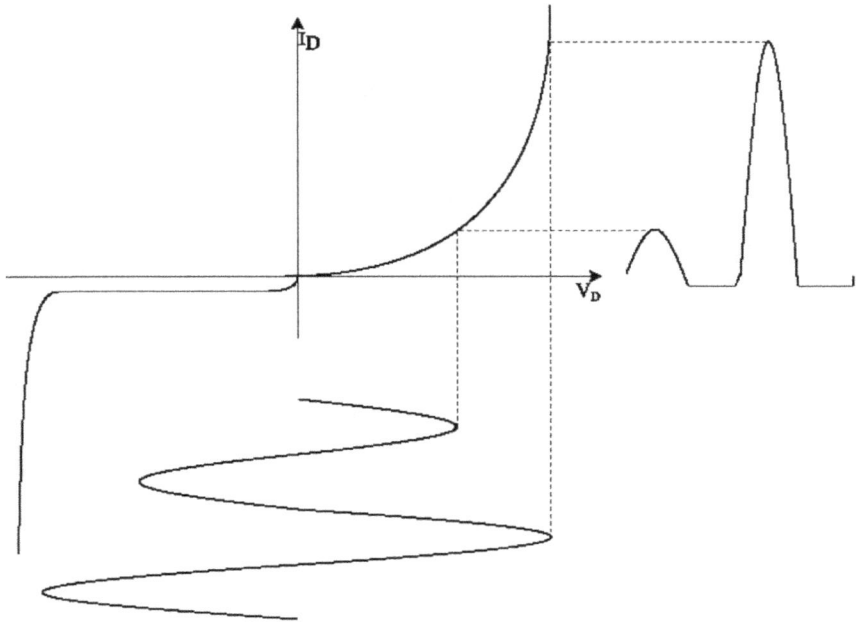

FIGURE 4.2   Typical diode I–V curve.

This characteristic can be used to improve the rectenna's performance, as long as the amplitude of the input signal generated by the antenna does not reach the diode's breakdown voltage. In order to have higher output signals in the antenna, its impedance should be as high as possible.

$$I_D = I_S * \left(e^{\frac{v}{n^* V_T}} - 1 \right) \qquad (4.2)$$

The antenna, which can be represented as shown in Figure 4.3, absorbs the electrons from the RF microwave signal, sent from the transmitter, and generates a current. With this, it is possible to perform a theoretical analysis on how the radiative impedance from the antenna influences the RF-DC's efficiency. Through the analysis of the rectenna with a simple single shunt diode as the rectifier (Figure 4.4) and using Equations (4.3–4.5), it is possible to compute the efficiency when the radiative impedance varies. As is demonstrated in Figure 4.5, when the antenna's impedance increases, the efficiency also increases, tending, in this case, to 45%. To complement this study the output resistor was also changed, providing the

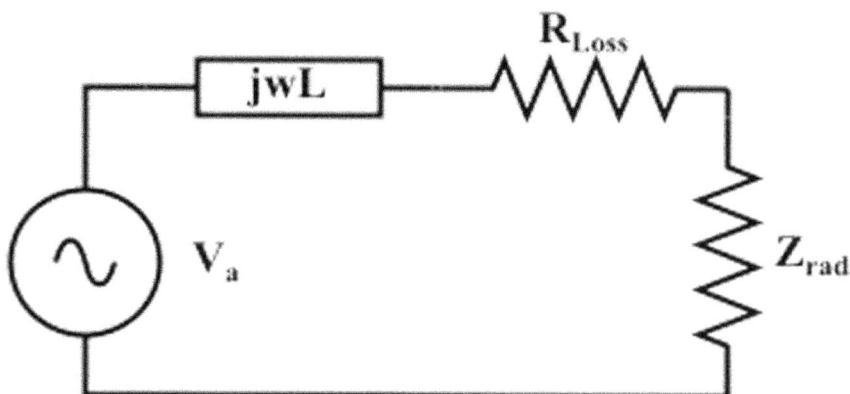

FIGURE 4.3   Electrical model of an antenna.

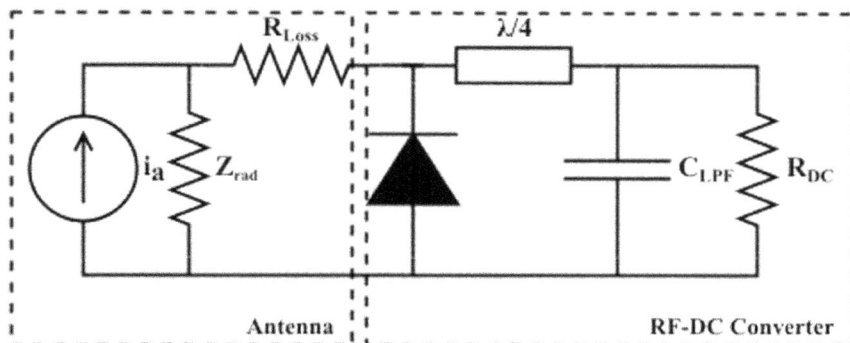

FIGURE 4.4   Rectenna's circuit schematic.

knowledge that, as expected, the diode's ports are interconnected and when one port's characteristics change, the other side is also affected. In Ref. [20], the importance of finding the best load value in an RF-DC converter is described (Figure 4.6). In Figure 4.7, the impacts of the impedance of the diode's two ports can be observed.

$$R_{out} = \frac{V_{R_{DC}}}{I_{R_{DC}}} \tag{4.3}$$

$$P_{out} = \frac{V_{R_{DC}}^2}{R_{out}} \tag{4.4}$$

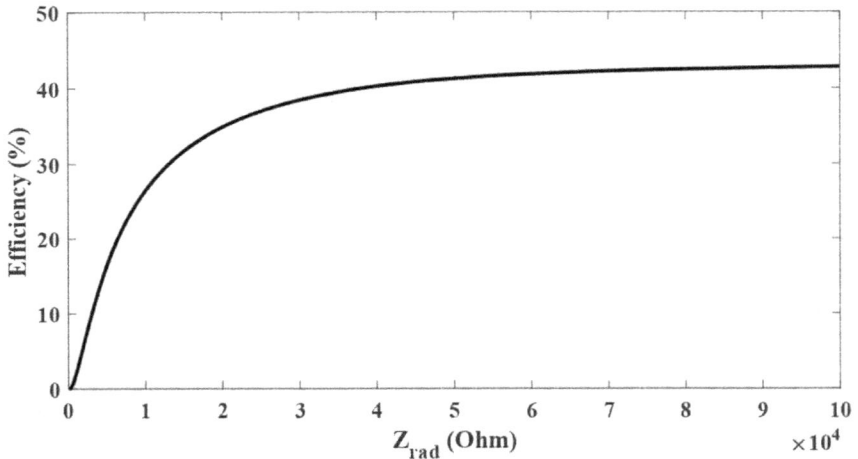

FIGURE 4.5 Rectenna's efficiency with the antenna's impedance variation.

$$Eff = 100 * \frac{P_{out}}{P_{in}} \qquad (4.5)$$

4.2.1.1.2 Device-level Theoretical Considerations: Load/Source Pull Simulation In order to understand how the source and load impedance influences a component or circuit, a load/source pull analysis is normally used. This type of analysis consists of varying the impedance seen by the component while computing the desired output results.

As the main objective is to perform an analysis of the correlation between the antenna's impedance and the rectifier's correlation performance, the load/source pull simulation is one of the most preferable simulations. In this case, the chosen diode to perform this analysis was the Skyworks' SMS7630-040LF Schottky diode, due to its fast recovery time and low power consumption. In Table 4.2 and Figure 4.8 the parameters used to configure the diode in the simulation software and the parasitic components added to the circuit are presented, with the intention of having the most reliable results possible.

To perform an accurate simulation and analysis, Keysight's Advanced Design System (ADS) was used due to its advanced simulation tools and configurable components. As shown in Figure 4.9, the diode was preceded and followed by impedance blocks which varied the impedance in the fundamental frequency (2.4 GHz) and its first three harmonics (4.8 GHz,

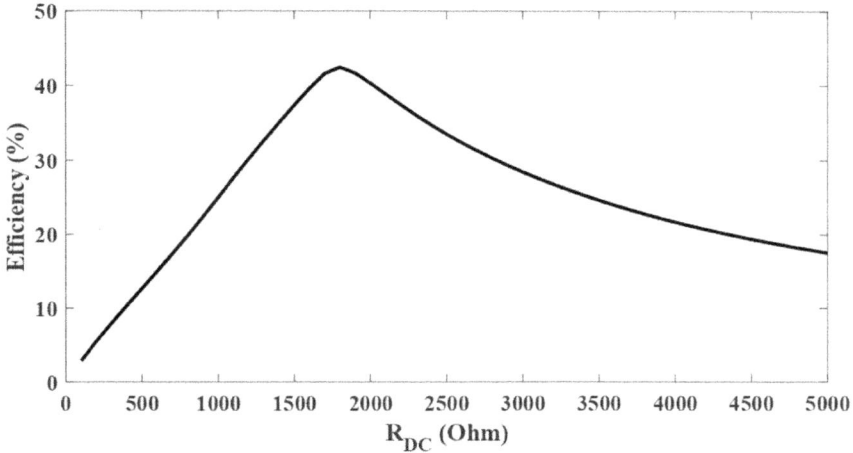

FIGURE 4.6   Rectenna's efficiency for several values of DC resistor.

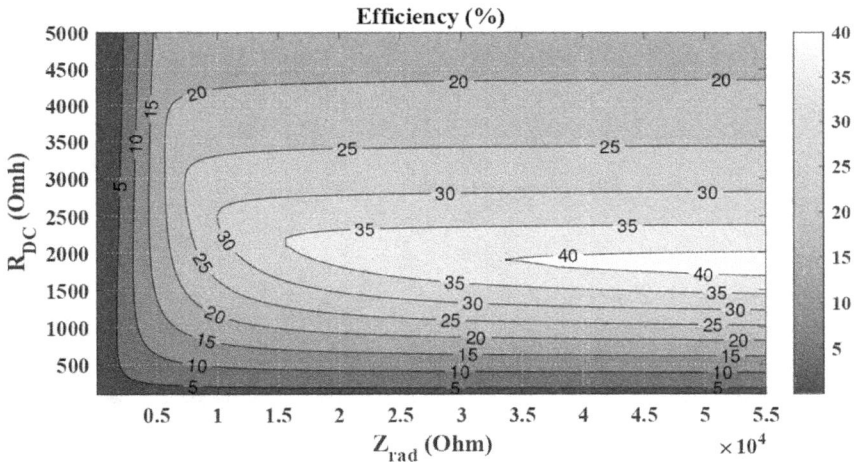

FIGURE 4.7   Rectenna's efficiency dependency with impedance.

7.2 GHz, and 9.6 GHz). To perform the load/source pull simulation the Harmonic Balance (HB) simulation was used alongside the Parameter Sweep tool in order to calculate the output efficiency of the overall converter for the many impedance combinations.

As it is possible to observe in Figure 4.10, the diode's rectifying behavior is improved if the impedance seen from the source at the fundamental frequency is as high as possible, being almost an open circuit (1861.70

TABLE 4.2   Diode SMS7630 Spice Parameters

Parameter	Value
Is	5 μA
Rs	20 Ω
N	1.05
Tt	1E-11 s
Cjo	0.14 pF
Vj	0.34 V
M	0.4
Fc	0.5
Bv	2 V
Ibv	1E-4 A
Xti	2
Eg	0.69

Source: [8].

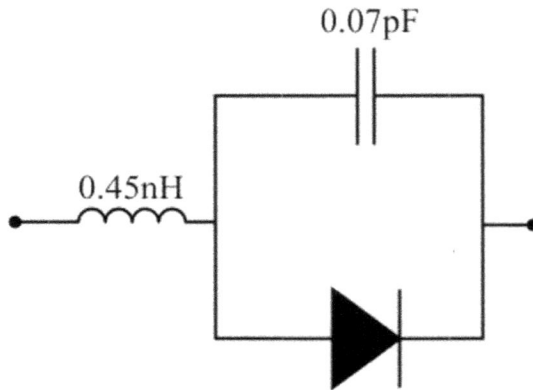

FIGURE 4.8   Diode SMS7630 electrical model with parasitic elements.

Source: [8].

+ j204.70), achieving in this case approximately 72% efficiency. When it comes to the harmonics, the values do not follow a specific rule, the second harmonic should be 2.62E-3 − j11 and the third harmonic should be 11.20E-3 − j93.50, as it is possible to see in Figure 4.11.

On the load side of the diode, there is still the need for harmonic rejection, as it is possible to observe in Figure 4.12, in which the optimum impedance is 0.15 + j397.90 at the fundamental frequency.

Through the analysis of the source pull simulation, it is possible to conclude that the diode achieves its pick performance when the source's

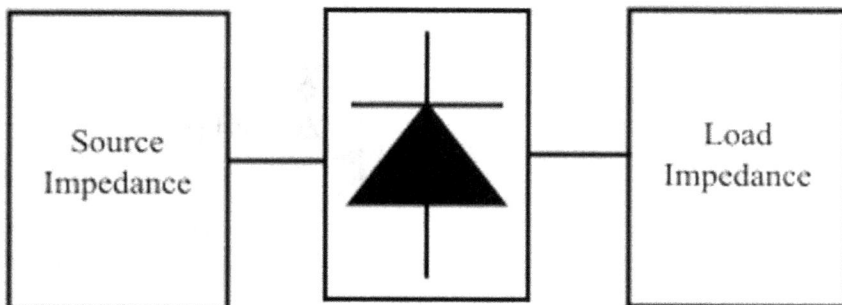

FIGURE 4.9   Schematic of the load/source pull simulation.

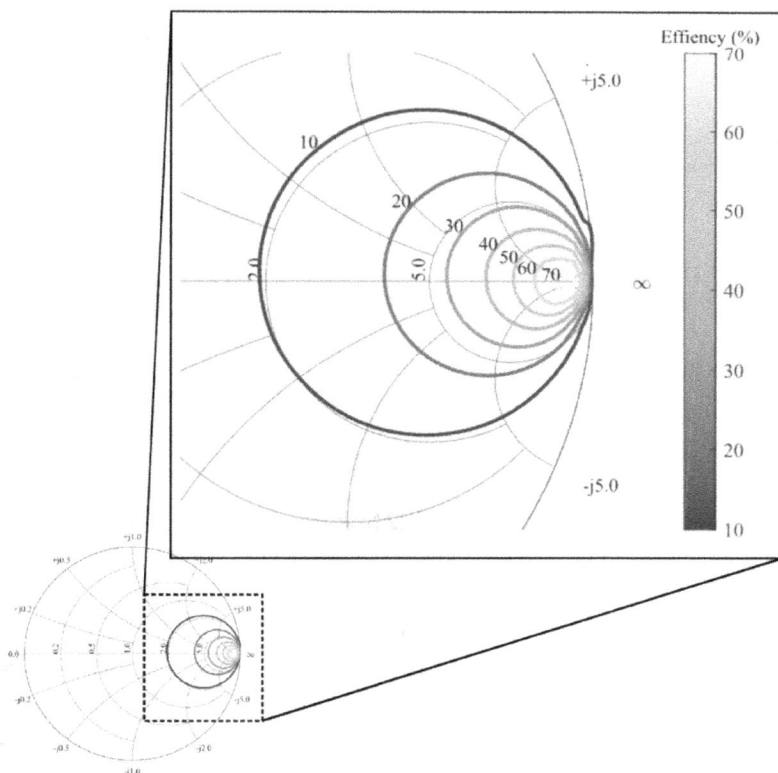

FIGURE 4.10   Source-pull simulation results.

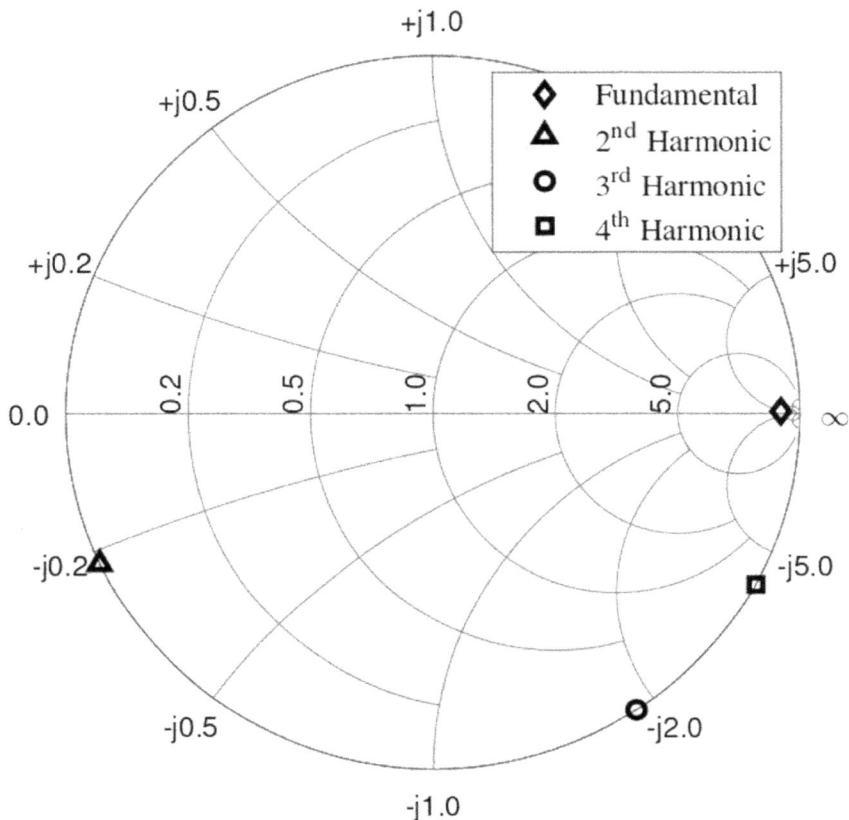

FIGURE 4.11   Source-pull harmonic simulation results for –5 dBm.

impedance is very high. In Section .4.2.1.2 this analysis will be implemented, simulated, and validated.

4.2.1.2 RF-DC Converter's Designing Procedures

In order to demonstrate which input impedance improves the diode's rectifying behavior and allows the achievement of the best RF-DC efficiency possible, the antenna was replaced by an impedance block and two circuits with a chosen frequency of 2.4 GHz were implemented experimentally. The first one follows the traditional approach of matching the impedance in both sides of the diode, through the complex conjugate method, while the second one implements the theory presented by the authors: the implementation of an impedance block with the impedance obtained through the source pull analysis, i.e. the source impedance is very high, almost

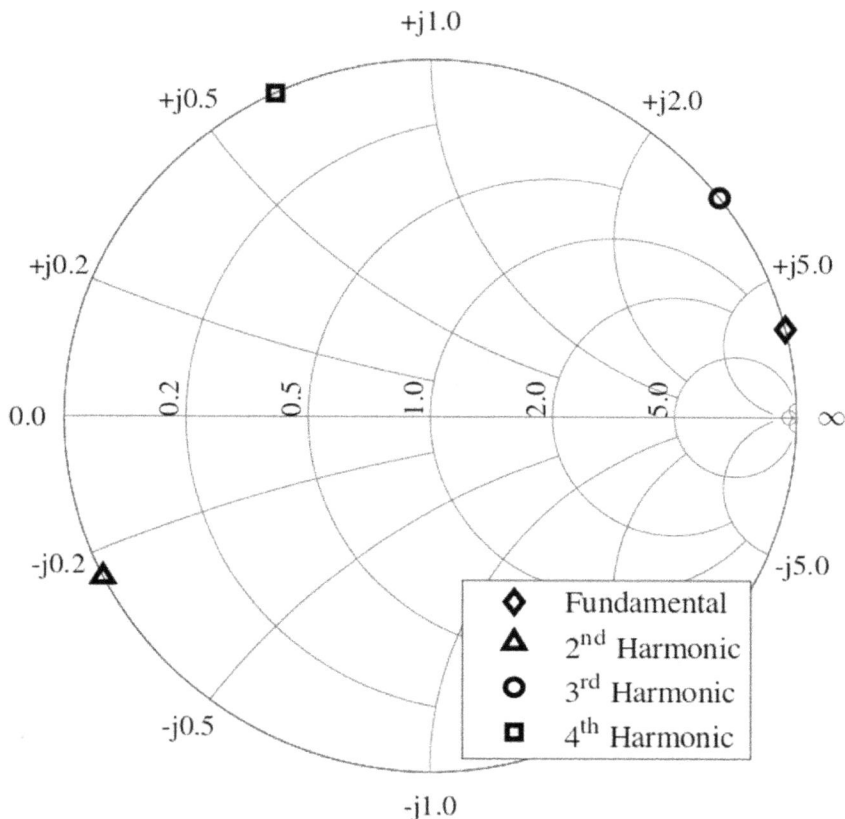

FIGURE 4.12   Load-pull simulation results for –5 dBm.

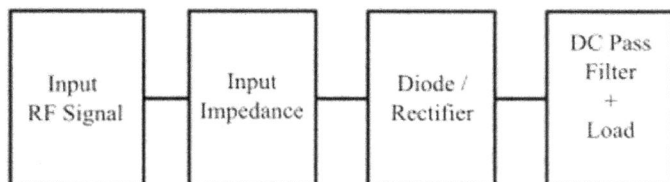

FIGURE 4.13   RF-DC converter's block diagram.

creating an open circuit. Both RF-DC converters designed in this section follow the block schematic of Figure 4.13, in which, firstly, the DC pass filter is a simple RC filter. Secondly, the chosen rectifier's architecture is the single-shunt diode converter, which is one of the simplest rectifiers. And finally, the input impedance block is either the match impedance or

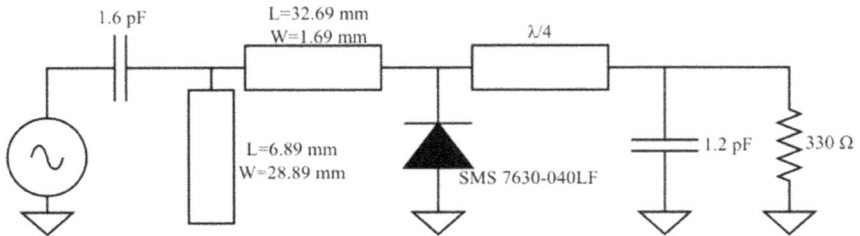

FIGURE 4.14   Matched impedance RF-DC converter schematic.

high-impedance block. As the objective of this chapter is to analyze how the input impedance affects the diode's behavior, no improvements were made to the converters through harmonic rejections.

4.2.1.2.1 Match Impedance Approach In the first design an RF-DC converter was implemented with the diode's complex conjugate impedance on either side, i.e. using a match impedance approach. To achieve the best results possible the goals imposed on the design were to obtain the best matching impedance through the complex conjugate while achieving the best efficiency possible.

As it is possible to observe in Figure 4.14, the converter's circuit is very simple, being composed of two stubs as the matching impedance block, a shunt SMS 7630-040LF Schottky diode, and a parallel capacitor filtering the rectified signal. In order to achieve the imposed goals, the length and width values of the matching impedance stubs were varied, as well as the capacitor and resistor values. With these requirements the circuit designed can be found in Figure 4.16 with an efficiency of 5% at 5 dBm and a pick efficiency of 8.1% at 13 dBm, and the source impedance is the complex conjugate of the load impedance seen by the diode, as it is possible to see in Figure 4.15.

4.2.1.2.2 Proposed High-impedance Approach The proposed converter approach uses high impedance in order improve the circuit's efficiency. The circuit can still be divided into the blocks shown in Figure 4.13, and, in this case, the input impedance block transforms the input impedance, which is normally 50 Ω, in high impedance. Using the same approach as presented, it was possible to design an RF-DC converter with the same input impedance as the one obtained in the source-pull simulation. The main goals of this design were to have the input impedance in the fundamental frequency seen by the diode equal to the one obtained in the source-pull simulation. And secondly, the circuit's efficiency should be as high as possible.

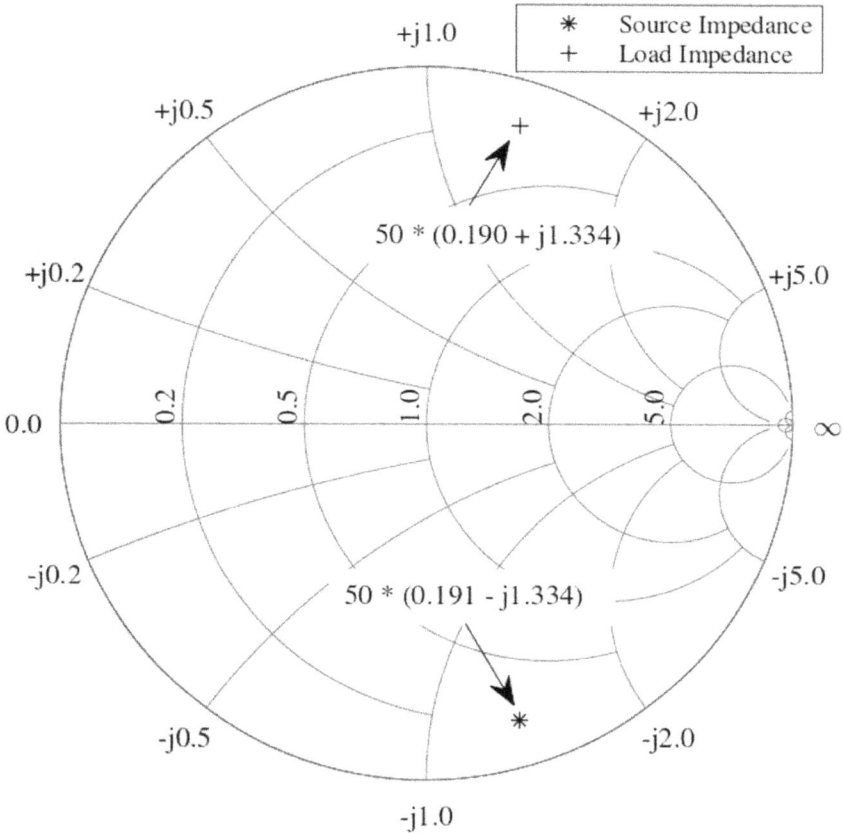

FIGURE 4.15   Simulated source and load impedance of the diode.

As shown in Figure 4.17, the circuit is very similar to the one designed in the previous section, with only one main difference, which is the use of more stub phases in the input impedance block, in order to obtain the most similar impedance results to the source pull simulation. With the purpose of achieving the imposed goals, the length and width values of the input block's stubs were varied, and the capacitors and resistor were also varied. With this, it was possible to design the circuit in Figure 4.17 with an efficiency of 35.6% at 0 dBm and 21.3% at 5 dBm (Figure 4.19), and the input impedance is almost an open circuit at the fundamental frequency (Figure 4.18), which is very close to the values obtained in the source pull simulation. When comparing this simulation with the source pull results, it is possible to conclude that it was not possible to achieve the same efficiency as the source pull, due to the impedance at

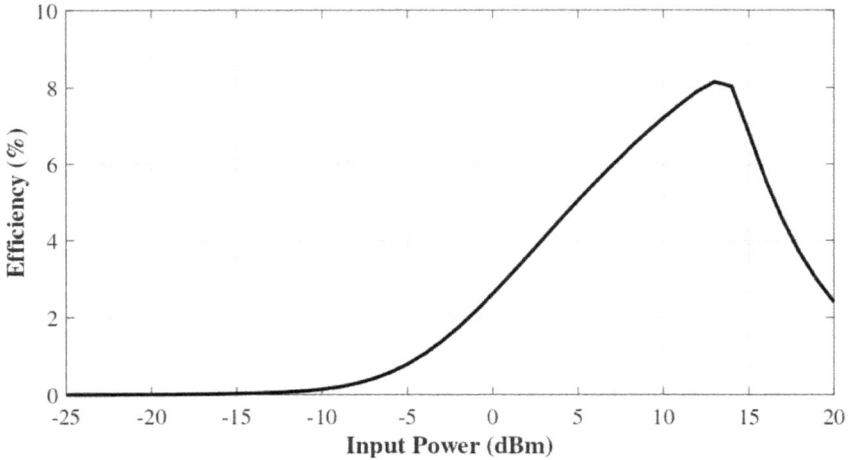

FIGURE 4.16 Matched impedance RF-DC converter simulated efficiency.

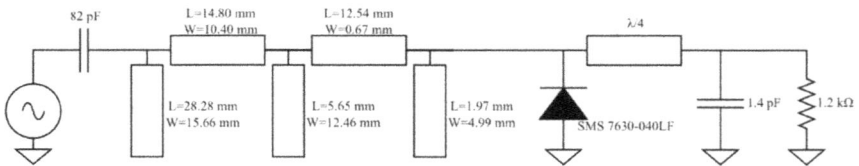

FIGURE 4.17 High input impedance RF-DC converter: schematic.

the fundamental frequency not being as high as required and the lack of harmonic rejection, which could possibly be corrected by having more stub phases. Nevertheless, this circuit had better results than the match impedance approach.

When comparing the simulation results, it is already possible to infer that there is a major difference when using high impedance in the converter's input instead of matching it with the complex conjugate. In this case, the results are more than four times higher if the input impedance is almost an open circuit, and the pick efficiency in the high-impedance circuit occurs at a much lower input power, whereas the match impedance circuit needs higher input power.

4.2.1.2.3 Experimental Results With the objective of testing experimentally the circuits simulated and confirming their results, the Isola's Astra 3 substrate with a thickness of 0.762 mm, a dielectric constant of 3, and a dissipation factor of 0.0017 was chosen. Figures 4.20 and 4.21 show the

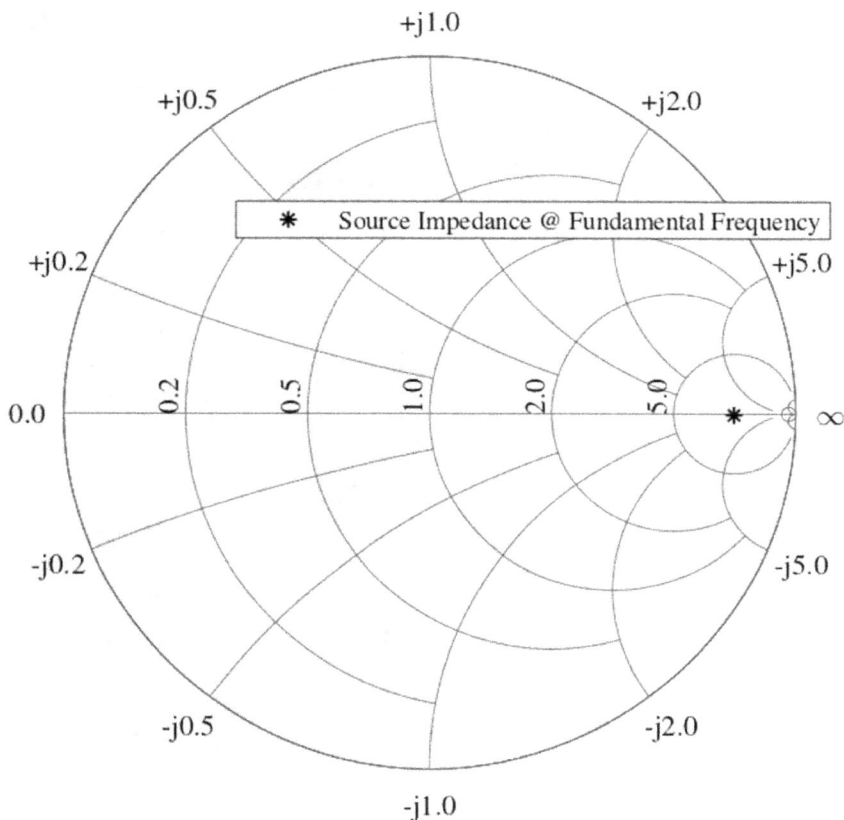

FIGURE 4.18   Simulated source impedance of the diode.

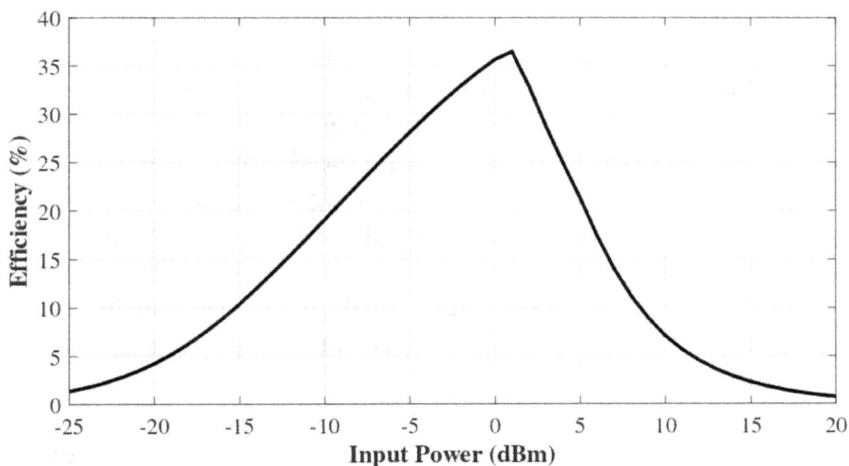

FIGURE 4.19   High input impedance RF-DC converter simulated efficiency.

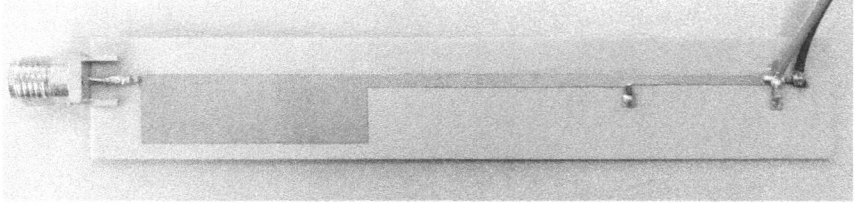

FIGURE 4.20   Manufactured PCB of the matching impedance converter.

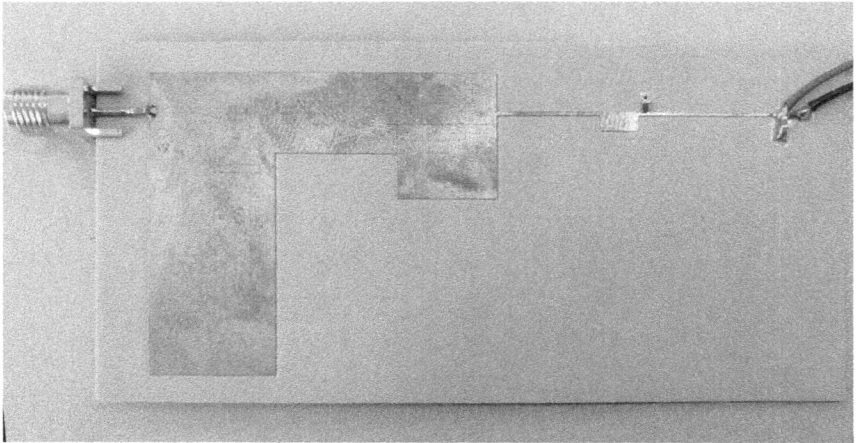

FIGURE 4.21   Manufactured PCB of the high impedance converter.

manufactured printed circuit boards (PCBs) for the matched impedance converter and high input impedance converter, respectively.

The setup used to measure the output results from the built PCBs was composed by the Keysight's PSG Vector Signal Generator E8267D to supply the input signal to the converters, and the multimeter 34461A, also from Keysight, to measure the output voltage. Each piece of equipment was connected to MATLAB via LAN with the objective of computing the RF-DC conversion's efficiency and automating the measurement process. The input signal consisted of a sinusoidal wave, in which the power was made to vary from −20 dBm up to 20 dBm and the frequency from 2.3 GHz to 2.5 GHz, with the objective of analyzing its behavior in low and high input power and how a slight change in the frequency influences the conversion.

As it is possible to observe in Figure 4.22, the efficiency of the matched impedance rectifier is very low, tending to the 3% marker. The discrepancy

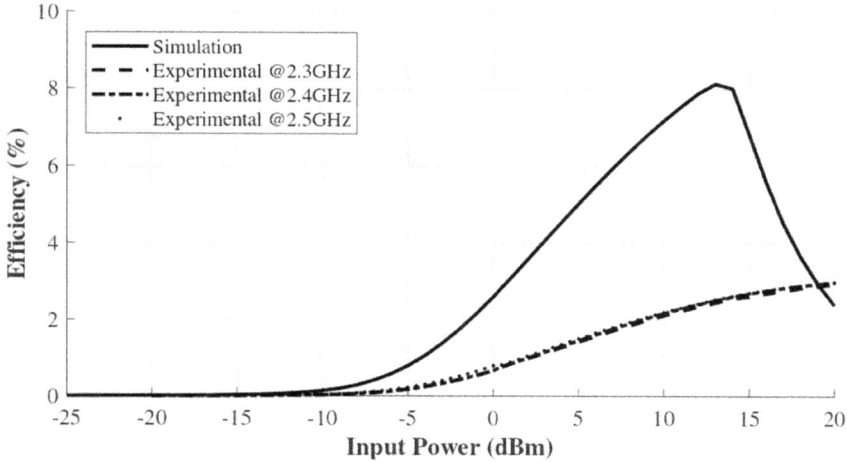

FIGURE 4.22   Matched impedance RF-DC converter efficiency comparison.

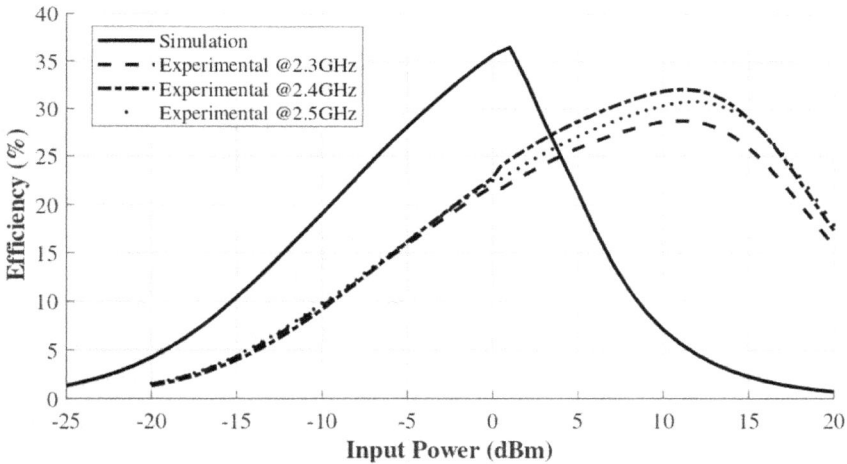

FIGURE 4.23   High input impedance RF-DC converter efficiency comparison.

between the experimental and simulated results could be due to the fabrication process. By analyzing the results with frequency variation, the results are very similar, meaning that the converter is very stable with slight changes in the frequency. Figure 4.23 presents the experimental results of the high input impedance rectifier, which present a slightly worse efficiency when compared to the simulated results, but nevertheless, the pick efficiency is still 33% at 11 dBm input power and at 5 dBm the converter has an efficiency of 28.5%. This converter presents a very similar performance

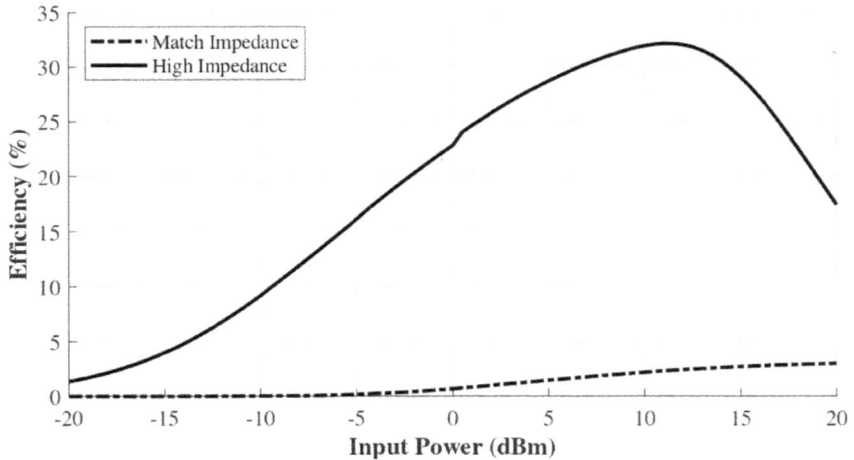

FIGURE 4.24 Comparison of the efficiency between the two produced circuits.

with the frequency variation as the previous one when the input power was under 0 dBm.

Through the analysis of Figure 4.24, it is possible to conclude that the high input impedance approach can improve significantly the rectifying behavior of a diode. As explained in Section 4.2.1.1, for a given current, if the impedance is increased the voltage will also increase. Moreover, due to the characteristic curve of the diode, the higher the input current that reaches the diode the higher the voltage rectified by the diode.

4.2.1.3 Final Remarks Regarding the Source Pull Designing Approach

This work has provided an analysis of the impacts of the impedance in the diode's performance, through the simulation of the diode and actual experimental implementation of two converters. Theoretically, the diode should be exposed to a higher impedance from the source, in order to increase its voltage output signal. The source pull simulation also provided the same conclusion, in which the results from this simulation imposed almost an open circuit in the source.

Through the comparison of the converters with an input block with high impedance and matched impedance it was possible to conclude that the impedance chosen in the input block has a significant impact on the diode's behavior, being the most favorable choice using high impedance in the source.

When comparing the source pull simulation to the experimental results it is possible to observe that there is a decrease in the efficiency, which could be due to the fabrication process and the lack of harmonic rejection.

Due to the diode's characteristic curve being exponential, it grants the possibility of having a significant change in the diode's output signal with the smallest change in the received signal, so by using high impedance it is possible to improve the diode's behavior. As in a rectenna, the diode is directly connected to the antenna, and by Ohm's law it is possible to increase the antenna's output signal by increasing its impedance. In this way the signal that reaches the diode is higher than when the complex conjugate is used to match the diode's impedance.

4.2.2 Hybrid mm-wave GaAs RF-DC Converter

4.2.2.1 RF-DC Converter Circuit Design and Implementation

The system presented in this work is a hybrid solution combining an RF-DC converter chip, based on a two-stage Dickson charge pump/voltage multiplier [26], that was developed using a 100 μm GaAs technology foundry. However, the DC low-pass filter was developed externally to the chip, using a conventional microstrip substrate, as shown in Figure 4.25. The chip is composed of a matching network (values in Figure 4.25) that matches the input impedance to the diode's impedance. Then, the connection between

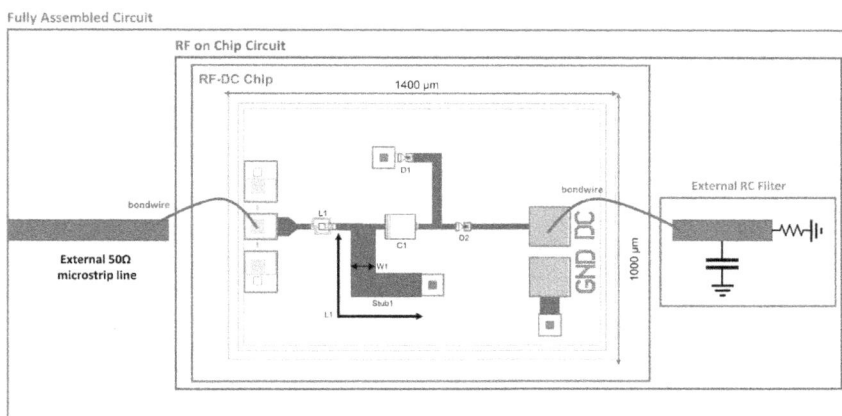

FIGURE 4.25 RF-DC block diagram circuit for different scenarios. L1 = 0.12 nH, C1 = 2.7 pF, L1 stub = 372 μm, W1 stub = 90 μm, WD1 = WD2 = 10 (N° of fingers). The area of the chip is 1400 × 1000 μm.

the chip and the external RC filter is made by a bondwire. This external RC filter allows tuning the circuit for different scenarios. For example, if it is intended to measure the chip using a probing station the RC network is not the same as if the circuit were connected to an antenna since the 50 Ω microstrip line and bondwire would influence the chip response. This enables optimization of the circuit for different scenarios but also different frequencies of operation. The substrate used was the Rogers RO4003C, with Er = 3.48 with h = 0.508 mm, and both parts of the circuit were optimized for a central frequency of 24 GHz and an input power of 10 dBm.

4.2.2.2 Measurement Setup and Results Discussion

As mentioned earlier, the external RC network objective allows the circuit to be evaluated in various scenarios. In this work, the scenarios that will be addressed are an RF on-chip, this is, using a probing station and another with a fully assembled circuit for different levels of input power and a wide range of frequencies (20–28 GHz). The measurement setup for both scenarios is similar. A Vector Signal Generator (VSG) E8267D from Keysight generates and sends a continuous wave (CW) to the circuit with a power sweep from –20 to 25 dBm of input power. If the measurement performed is on-chip, a probe-station (Cascademicrotech 9000) with Cascade Microtech probes with 150 μm of pitch is used to send the signal in the chip. Then a digital multimeter 34461A from Keysight is used to measure the output voltage of the rectifier. All processing and calculations are performed in MATLAB. Also, before the measurement, a power calibration is performed to assure the cables' losses and connectors that will be used to send the correct power to the circuit. To accomplish this, the power-meter NRP33SN from Rohde & Schwarz was used. In Figure 4.26, is illustrated a microscopic photograph of the chip and the prototypes produced for each scenario.

In Figure 4.27 a comparison between the results obtained in the simulations (where both the external networks and bond wires were considered) and the measured results for each prototype is provided. Analyzing this makes it possible to conclude that the results obtained in the measurements are close to the simulated results. However, a more detailed analysis will be performed in the following sections. Nevertheless, it was observed during the simulations that the diode model could have been better characterized above 25 dBm since the simulation curve in Figure 4.27 shows unusual behavior.

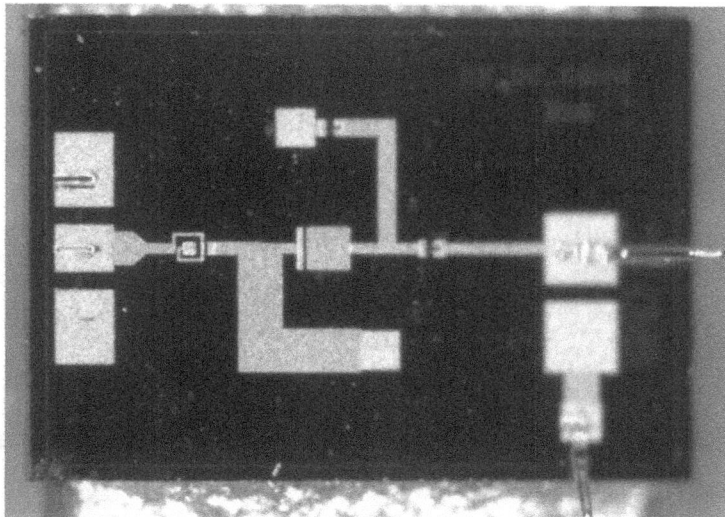

FIGURE 4.26 (a) Microscopic photograph of the chip; (b) on-chip scenario prototype; (c) fully assembled circuit prototype.

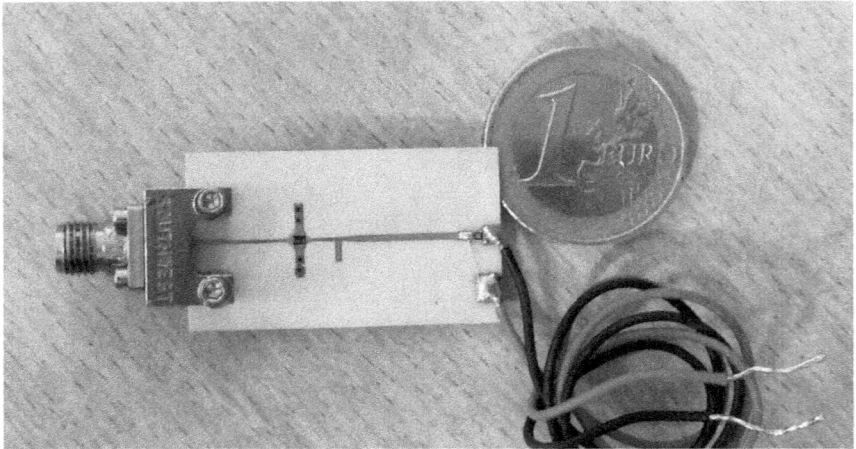

FIGURE 4.26 (Continued)

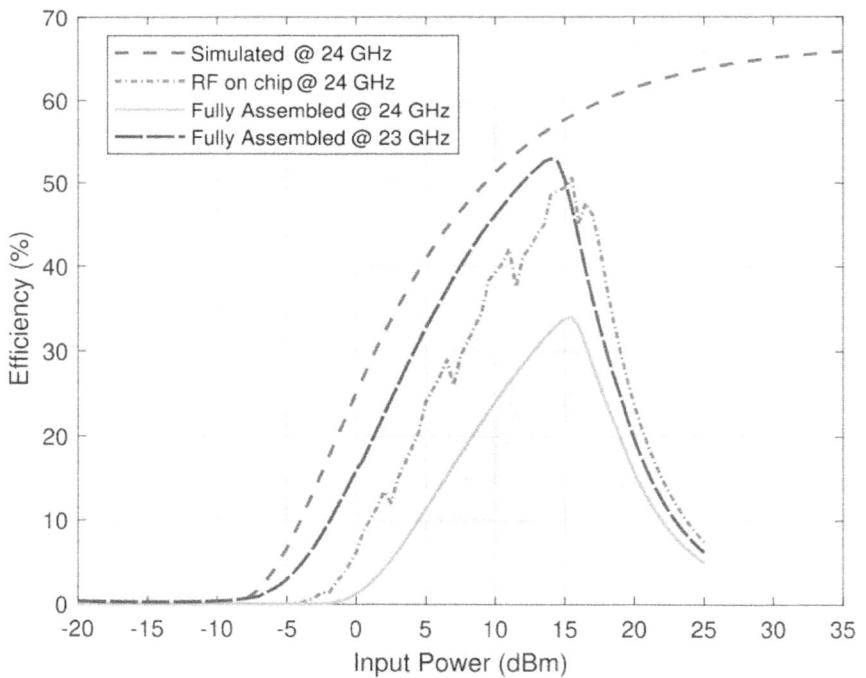

FIGURE 4.27 Comparison between simulated and measured results for different scenarios.

4.2.2.2.1 On-chip Measurements The on-chip measurement scenario (shown in Figure 4.28) is analyzed in this section. The external low-pass network was designed and optimized to achieve the maximum efficiency, considering that the feed was made directly in the chip. With this in mind, the external RC filter resulted in an open stub with an equivalent capacity of 163 fF and an optimal load of 1.96 kΩ. The results of these measurements are presented in Figure 4.28. In the several figures presented, it is possible to see that the efficiency curve shows some "steps" that are a result of the transition between the probe and the chip.

Regarding the results achieved, the efficiency is relatively lower than the simulated results. However, the circuit shows similar behavior for different loads but also for several frequencies. For example, at 25 GHz, the efficiency curve is almost equal to the 24 GHz. This shows that the circuit

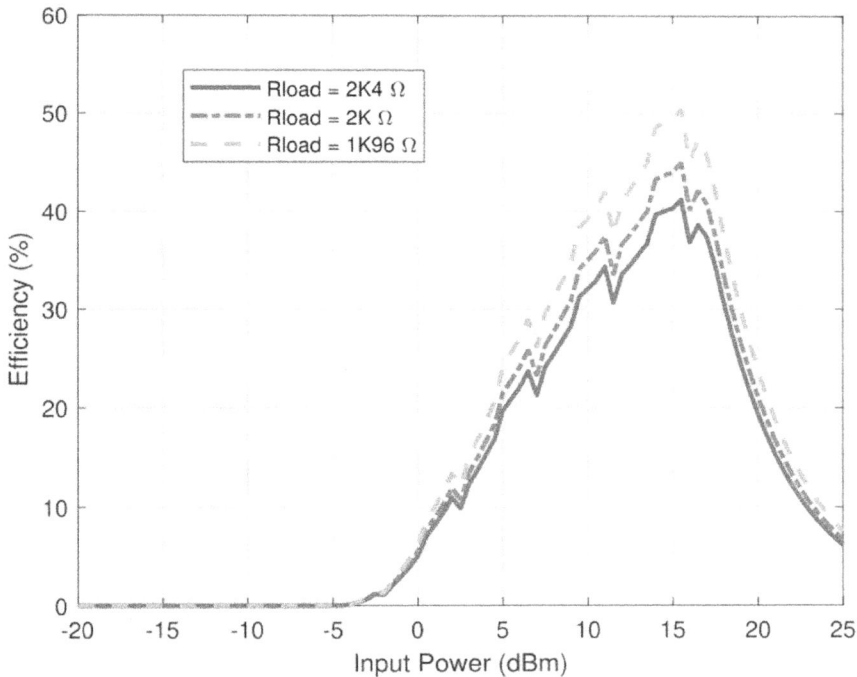

FIGURE 4.28   Results of on-chip scenario: (a) efficiency for different loads at 24 GHz; (b) efficiency for several frequencies for a 1K96 Ω load; (c) output voltage for different frequencies for a 1K96 Ω load.

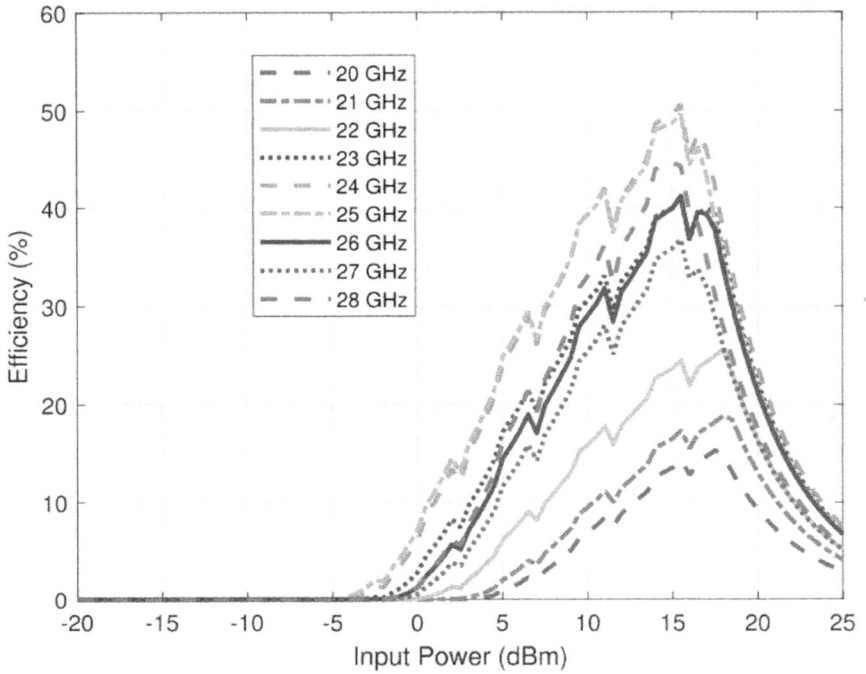

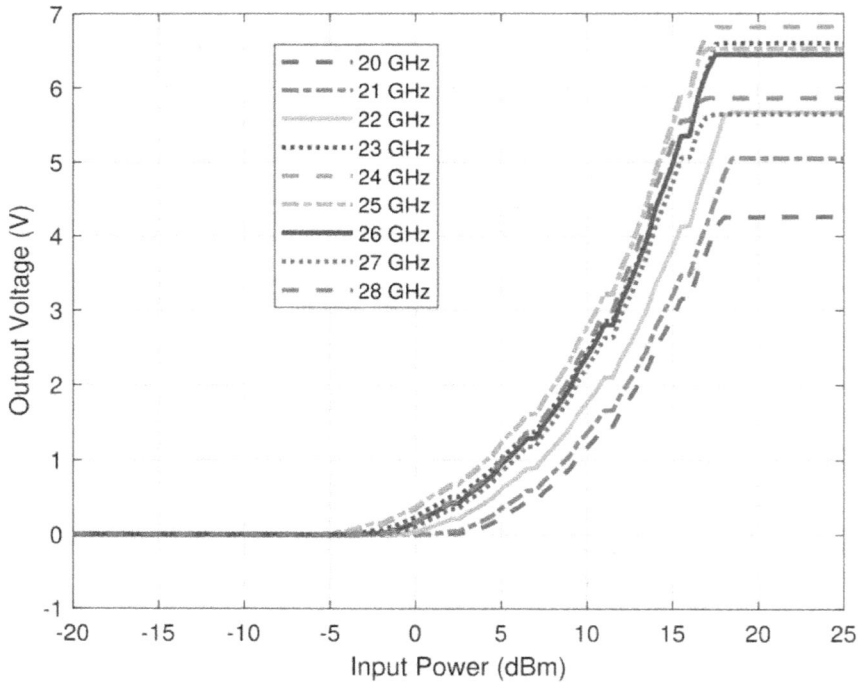

FIGURE 4.28   (Continued)

has an excellent performance considering different variables (loads and frequencies).

4.2.2.2.2 Fully Assembled Circuit Measurements For this case, where the fully assembled circuit is considered (shown in Figure 4.29), the methodology followed is the same as that presented in the previous section. The external RC filter comprises an open stub with an equivalent capacity of 584 fF and an optimum load of 3.3 kΩ. The results of these measurements can be seen in Figure 4.29. By observing Figure 4.27, a shift/deviation occurred since the maximum efficiency appears at 23 GHz and not at 24 GHz as intended. This can be explained by the production and assembling process since the poor characterization of bondwires is known about. Nevertheless, the results at 24 GHz are suitable, as for other frequencies. At 23 GHz, by varying the load, the behavior is similar to the on-chip case. With this, we can conclude that both prototypes present an excellent performance in the scenarios to which they were subjected.

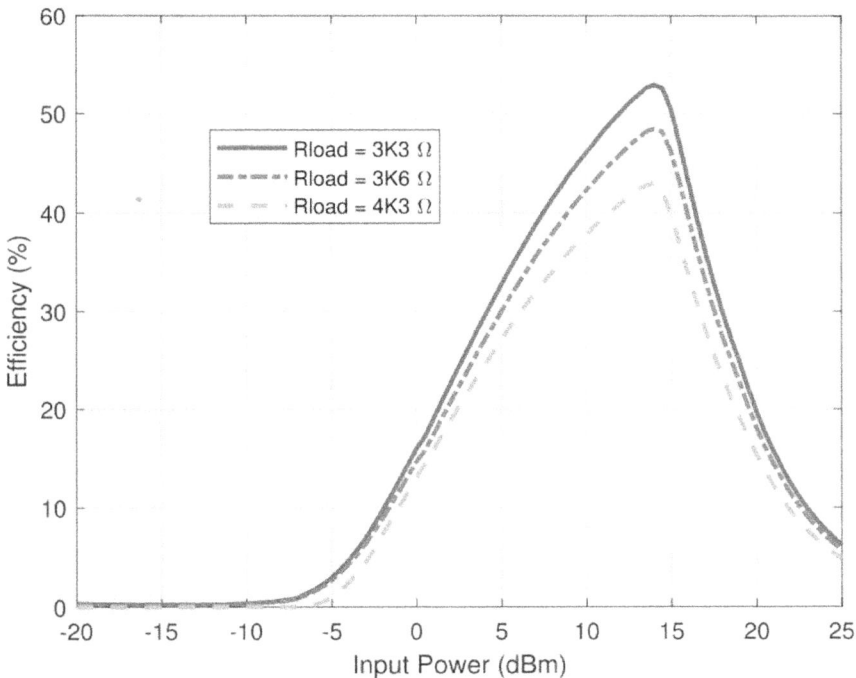

FIGURE 4.29 Results of the fully assembled circuit: (a) efficiency for different loads at 23 GHz; (b) efficiency for several frequencies for a 3K3 Ω load; (c) output voltage for different frequencies for a 3K3 Ω load.

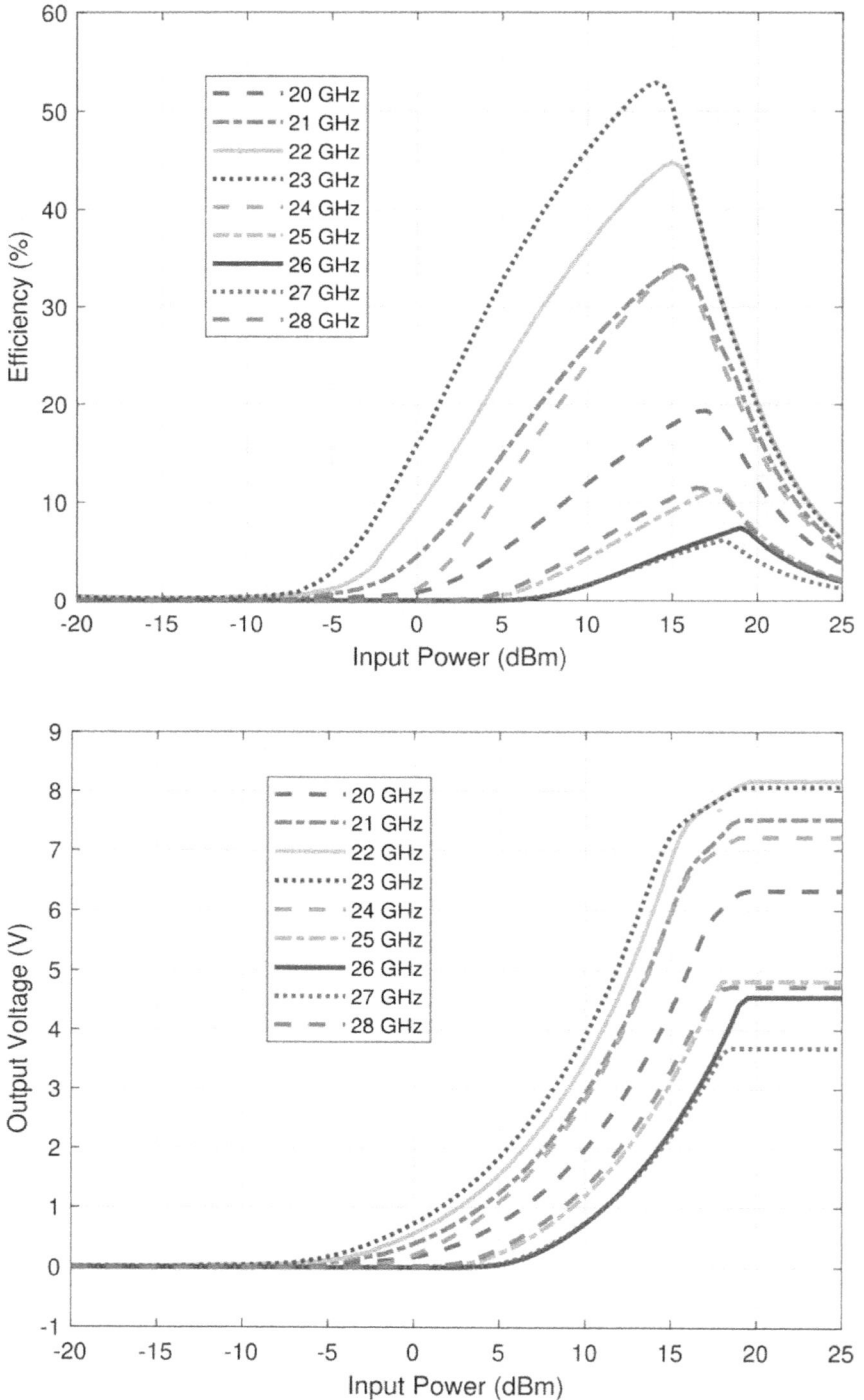

FIGURE 4.29 (Continued)

TABLE 4.3 Performance Summary and Comparison

References	[21]	[22]	[23]	[24]	[25]	This work
Technology	GaAs chip	Discrete	GaAs chip	Discrete inkjet print	Discrete	Hybrid (GaAs chip + microstrip)
Frequency	5.8 GHz	24 GHz	24 GHz	24 GHz	35 GHz	24 GHz
Efficiency	35% @15 dBm	42% @ 13 dBm	27% @ 15 dBm	35% @ 15 dBm	40% @ 13 dBm	54% @ 13 dBm
	67% @ 30 dBm		48% @ 23 dBm		60% @ 21 dBm	51% @ 15 dBm
Output voltage	N.R.	0.9 V	N.R.	N.R.	N.R.	8V

4.3 CONCLUSION

The design, implementation, and measurement of a mm-wave hybrid RF-DC converter was presented in this chapter. The proposed circuit and the scenarios considered were described, and the respective considerations for producing the respective prototypes were also explained. A full characterization was then performed for each of the prototypes produced as it was measured for different loads and multiple frequency values with an input power sweep from –20 to 25 dBm. For the case of on-chip measurement, it achieved 51% efficiency for an input power of 15 dBm. In the fully assembled prototype, some deviation was found, and the peak of efficiency appears at 23 GHz with 54% efficiency with 13 dBm of input power. Nevertheless, both circuits presented promising results at other frequencies. Finally, a high output voltage is also achieved, with a maximum value of 8 V, considering a load of 1.96 kΩ. Table 4.3 presents a comparison between these circuits and those from other published works.

ACKNOWLEDGMENT

We acknowledge Helena Ribeiro, Diogo Matos, and Ricardo Correia for their efforts in building the circuits and solutions presented in this chapter. The work was published in high-quality papers, and in this chapter, there is a summary of previously presented activities.

REFERENCES

[1] W. C. Brown, "The History of Power Transmission by Radio Waves," *IEEE Transactions on Microwave Theory and Techniques*, vol. 32, no. 9, pp. 1230–1242, 1984.

[2] T. Lee, "Global Internet Access from Space for Humanitarian Applications," in *IEEE International Frequency Control Symposium (IFCS)*, 2016.

[3] N. A. Pantazis, S. A. Nikolidakis and D. D. Vergados, "Energy-Efficient Routing Protocols in Wireless Sensor Networks: A Survey," in *IEEE Communications Surveys and Tutorials*, vol. 15, no. 2, pp. 551–591, Second Quarter 2013. doi: 10.1109/SURV.2012.062612.00084

[4] E. D. N. Annex, "*Energy Efficiency of the Internet of Things*", 2016.

[5] S. Yinbiao et al., "Internet of Things: Wireless Sensor Networks," *IEC White Paper*, 2014.

[6] S. D. Assimonis, S. Daskalakis and A. Bletsas, "Sensitive and Efficient RF Harvesting Supply for Batteryless Backscatter Sensor Networks,"

in *IEEE Transactions on Microwave Theory and Techniques*, vol. 64, no. 4, pp. 1327–1338, April 2016. doi: 10.1109/TMTT.2016.2533619

[7] H. J. Visser, "A Brief History of Radiative Wireless Power Transfer," *2017 11th European Conference on Antennas and Propagation (EUCAP)*, pp. 327–330, 2017. doi: 10.23919/EuCAP.2017.7928700

[8] Skyworks Solutions, Inc., *Data Sheet Surface Mount Mixer and Detector Schottky Diodes*, March 13, 2013.

[9] Z. Popović et al., "Scalable RF Energy Harvesting," in *IEEE Transactions on Microwave Theory and Techniques*, vol. 62, no. 4, pp. 1046–1056, April 2014. doi: 10.1109/TMTT.2014.2300840

[10] S. Korhummel, D. G. Kuester and Z. Popović, "A Harmonically-Terminated Two-Gram Low-Power Rectenna on a Flexible Substrate," *2013 IEEE Wireless Power Transfer (WPT)*, pp. 119–122, 2013. doi: 10.1109/WPT.2013.6556897

[11] S. D. Joseph, S. S. H. Hsu, A. Alieldin, C. Song, Y. Liu and Y. Huang, "High-Power Wire Bonded GaN Rectifier for Wireless Power Transmission," *IEEE Access*, vol. 8, pp. 82035–82041. 2020. https://doi.org/10.1109/ACCESS.2020.2991102

[12] S. U. Din, Q. Khan, F. Rehman and R. Akmeliawanti, "A Comparative Experimental Study of Robust Sliding Mode Control Strategies for Underactuated Systems," in *IEEE Access*, vol. 5, pp. 10068–10080, 2017. doi: 10.1109/ACCESS.2017.2712261

[13] X. Li, L. Yang and L. Huang, "Novel Design of 2.45-GHz Rectenna Element and Array for Wireless Power Transmission," in *IEEE Access*, vol. 7, pp. 28356–28362, 2019. doi: 10.1109/ACCESS.2019.2900329

[14] M. Wagih, A. S. Weddell and S. Beeby, "Meshed High-Impedance Matching Network-Free Rectenna Optimized for Additive Manufacturing," in *IEEE Open Journal of Antennas and Propagation*, vol. 1, pp. 615–626, 2020. doi: 10.1109/OJAP.2020.3038001

[15] C. Song et al., "Matching Network Elimination in Broadband Rectennas for High-Efficiency Wireless Power Transfer and Energy Harvesting," in *IEEE Transactions on Industrial Electronics*, vol. 64, no. 5, pp. 3950–3961, May 2017. doi: 10.1109/TIE.2016.2645505

[16] N. Sakai, K. Noguchi and K. Itoh, "A 5.8-GHz Band Highly Efficient 1-W Rectenna With Short-Stub-Connected High-Impedance Dipole Antenna," in *IEEE Transactions on Microwave Theory and Techniques*, vol. 69, no. 7, pp. 3558–3566, July 2021. doi: 10.1109/TMTT.2021.3074592

[17] N. Shinohara and Y. Zhou, "Development of Rectenna with High Impedance and High Q antenna," *2014 Asia-Pacific Microwave Conference*, pp. 600–602, 2014.

[18] M. Wagih, A. S. Weddell and S. Beeby, "High-Efficiency Sub-1 GHz Flexible Compact Rectenna Based on Parametric Antenna-Rectifier Co-Design," 2020 *IEEE/MTT-S International Microwave Symposium (IMS)*, pp. 1066–1069, 2020. doi: 10.1109/IMS30576.2020.9223796

[19] M. Ito et al., "High Efficient Bridge Rectifiers in 100MHz and 2.4GHz Bands," 2014 *IEEE Wireless Power Transfer Conference*, pp. 64–67, 2014. doi: 10.1109/WPT.2014.6839595

[20] F. Bolos, J. Blanco, A. Collado and A. Georgiadis, "RF Energy Harvesting From Multi-Tone and Digitally Modulated Signals," in *IEEE Transactions on Microwave Theory and Techniques*, vol. 64, no. 6, pp. 1918–1927, June 2016. doi: 10.1109/TMTT.2016.2561923

[21] F. Cheng, H. Ren, W. Jiang, M. Zhang, B. Zhang and K. Huang, "C-Band GaAs MMIC Rectifier for Wireless Power Transmission," in 2018 *IEEE Asia-Pacific Conference on Antennas and Propagation (APCAP)*, pp. 306–307, 2018.

[22] S. Ladan, A. B. Guntupalli and K. Wu, "A High-Efficiency 24 GHz Rectenna Development towards Millimeter-Wave Energy Harvesting and Wireless Power Transmission," *IEEE Transactions on Circuits and Systems I: Regular Papers*, vol. 61, no. 12, pp. 3358–3366, 2014.

[23] K. Hatano, N. Shinohara, T. Seki and M. Kawashima, "Development of MMIC Rectenna at 24 GHz," in *2013 IEEE Radio and Wireless Symposium*, pp. 199–201, 2013.

[24] S. Daskalakis, J. Kimionis, J. Hester, A. Collado, M. M. Tentzeris and A. Georgiadis, "Inkjet Printed 24 GHz Rectenna on Paper for Millimeter Wave Identification and Wireless Power Transfer Applications," in *2017IEEE MTT-S International Microwave Workshop Series on Advanced Materials and Processes for RF and THz Applications (IMWS-AMP)*, pp. 1–3, 2017.

[25] T. Yoo and K. Chang, "Theoretical and Experimental Development of 10 and 35 GHz Rectennas," *IEEE Transactions on Microwave Theory and Techniques*, vol. 40, no. 6, pp. 1259–1266, 1992.

[26] C. R. Valenta and G. D. Durgin, "Harvesting Wireless Power: Survey of Energy-Harvester Conversion Efficiency in Far-Field, Wireless Power Transfer Systems," *IEEE Microwave Magazine*, vol. 15, no. 4, pp. 108–120, 2014.

CHAPTER 5

Optical WPT

Tomoyuki Miyamoto

5.1 INTRODUCTION TO WIRELESS OPTICAL POWER TRANSMISSION

The use of cables limits the installation of equipment, its equipment, and the expansion of applications and services. Wireless power transmission is expected to bring about significant changes in society. Although there are various wireless power transmission (WPT) methods, optical wireless power transmission (OWPT), which uses light as the energy medium, as shown in Figure 5.1, is expected to expand its applications due to its attractive features [1].

OWPT is expected to offer various advantages such as long-distance power transmission, miniaturization, and the absence of electromagnetic interference (EMI). This chapter describes the principles, features, challenges, and recent trends of OWPTs. The principles of key devices such as light sources, light receivers, and related functional modules are explained. It also explains the long-range nature of OWPT technology and issues such as efficiency limits and safety. In addition, the latest trends and future applications of OWPT are presented, with examples of its application in small terminals, consumer devices, and mobile devices such as drones and micro-vehicles.

DOI: 10.1201/9781003328636-5

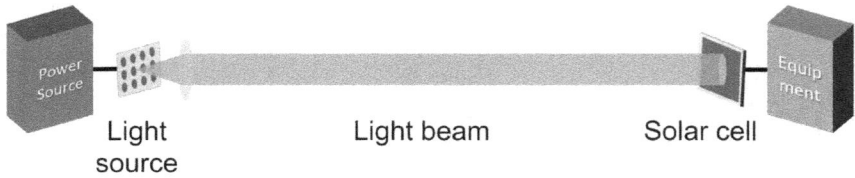

FIGURE 5.1   Basic configuration of OWPT.

5.1.1 Comparison of OWPT with other WPT Technologies

OWPT is a type of radiated wireless power transmission (WPT) using electromagnetic beams, similar to the RF (microwave) beam method. However, the frequency (wavelength) of light of about several hundred THz is about five orders of magnitude higher (shorter) than that of microwaves of several GHz, and not only is it possible to use small emitter and receiver devices corresponding to the wavelength, but also the diffraction phenomenon is very small even for narrow beam sizes, and the high beam collection rate can be maintained over long distances.

In addition, unlike other WPT methods that use AC circuits or RF radio waves, the light source and receiver operate on DC circuits, and there is no emission of RF radio waves. The light beam does not cause EMI to other equipment because the frequency of the light beam is far enough away from that of the circuit.

Because of these features, OWPT is a method that can power a variety of equipment wirelessly, opening the door to new equipment and new applications.

5.1.2 Is OWPT a New Technology?

At present, few people may know about the OWPT method. As already mentioned, it is a fascinating method, but has it been researched and developed (R&D) before? Or is it a new method?

The main devices used in the OWPT method are lasers as light sources and photovoltaic cells (solar cells) as light receivers. These devices were invented in the 1960s and 1950s, respectively. Therefore, the OWPT method may have been invented shortly after this period. In fact, several concepts of the OWPT method were reported around 1970 [2,3]. Since then, the research, development, and demonstration of several devices, systems, and principles have been reported, however there are not many in terms of the very broad field that is expected.

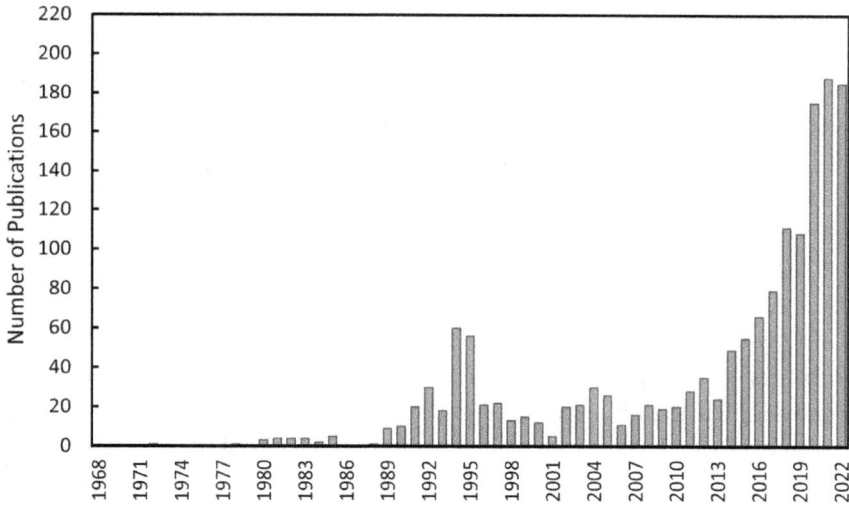

FIGURE 5.2   Number of publications related to OWPT. Nearly all data are based on the search keywords of "optical wireless power transmission" OR "optical wireless power transfer" OR "wireless optical power transmission" OR "laser wireless power transmission" OR "laser wireless power transfer" OR "wireless laser power transmission" OR "laser power beaming" in Google Scholar.

Figure 5.2 shows the number of OWPT-related articles, including both devices and systems. As the figure caption indicates, OWPT is referred to by various names, although "Optical Wireless Power Transmission" and "Laser Power Beaming" are well-known examples. For this reason, there are still many articles that are not included in this figure. It should also be noted that, since the data were mainly extracted from a web search system, it includes not only actual reports of OWPTs, but also mere mentions of keywords and completely unrelated articles.

From the 1980s through the 2000s, there were a few dozen articles per year. This number of articles is considered very small considering the wide range of relevant device types and possible applications. A similar search for several other fields yields 6,000 articles per year for "wireless power transmission," 20,000 articles per year for "energy harvesting," 20,000 articles per year for "IoT," and so on. A few dozen articles per year in the OWPT field seems far too few. Fortunately, however, the number of articles has been increasing since the mid-2010s and may reach a level comparable to the above fields in the next decade or so.

There has been some speculation as to why OWPT R&D has not been more widespread. The reasons are not clear, however (1) various WPT methods had already been developed, and a new method with a very different configuration and mechanism may have been considered unnecessary; (2) because it uses a light beam, the power transmission efficiency was considered very poor; (3) it was difficult to use high-power light sources that could handle the watt- to kilowatt-class required by many equipment; and (4) high-power beams, especially laser beams, are dangerous to use in areas where people are present.

Today, the situation is much better: (1) because the existing WPT has problems with long-distance power transmission and EMI, a new method is needed; (2) the efficiency of light sources and solar cells is improving; (3) high-power light sources are more readily available, and equipment that uses low power is increasing, so the overlap between the power that can be provided by the light source and the electrical power required in equipment is becoming close; and (4) advanced sensing technologies, such as high-performance cameras, and advanced recognition and control technologies, such as deep learning methods, are readily available to ensure safety. It is precisely this situation that will lead to further advances in technologies and applications for OWPT in the near future.

5.2 FUNDAMENTALS OF OWPT

5.2.1 Differences between OWPT and Solar Power Generation

Methods of generating electricity by irradiating solar cells with light have already been widely commercialized as photovoltaic power generation and indoor lighting power generation, as shown in Figure 5.3. However, these are essentially energy-harvesting methods. In the case of photovoltaic power generation, it can only be used at times and places where sunlight reaches. In addition, indoor lighting is not always on. Also, the total energy density of sunlight is 1 kW/m^2 (100 mW/cm^2) at the ground surface. This density (and light spectrum) is called the AM1.5 condition, where AM stands for "air mass" and the number is the thickness of the atmosphere through which it passes. AM0 is outside the atmosphere and AM1 is incident perpendicular to the Earth's surface. AM1.5 means passing through a layer of air 1.5 times the distance of AM1, which corresponds to conditions at slightly higher latitudes and is also used as a standard for the Earth's surface.

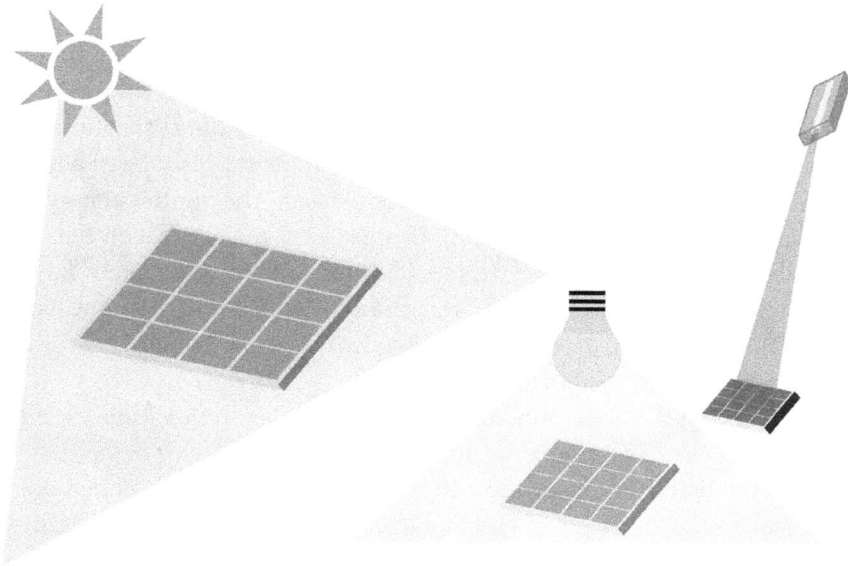

FIGURE 5.3   Power generation by sunlight, indoor lighting, and OWPT.

Since the photoelectric conversion efficiency of a typical solar cell is about 20%, a 10 cm × 10 cm solar cell would produce an output of 2 W of power under AM1.5 conditions. A 1 m × 1 m size would produce an output of 200 W. The output can be increased by increasing the area of the solar cell. It should be noted, however, that these outputs are not sufficient for various devices that would be equipped with solar cells. In the case of indoor lighting, the light output density is estimated to be as low as 1/100 or less of sunlight, thus further limiting its use.

OWPT is different from conventional photovoltaic power generation systems. OWPT uses a specific light source for power transmission. By providing a light source, the system can be supplied with the required power, power density, and appropriate light spectrum at any time.

5.2.2 Sunlight and Monochromatic Light Irradiation of Solar Cells

Sunlight has a broad light spectrum ranging from ultraviolet to infrared. The peak wavelength is near green (wavelength 550 nm), but the full width at half maximum (FWHM) of the spectrum is $\Delta\lambda = 500$ nm, which is as wide as $\Delta E = 2$ eV when considered in terms of photon energy. Because of this broad sunlight spectrum, conventional solar cells have a photoelectric conversion efficiency of about 20% (30% even under ideal conditions). The reason for this low efficiency is that low photon energies, i.e. wavelengths longer than the bandgap (absorption edge energy) of the solar cell, are not absorbed. In addition, although large photon energies are absorbed, the carriers generated by the absorption transition to the absorption edge energy in a very short time, and this transition generates a large amount of thermal energy which is the difference between the photon energy and the bandgap. Considering the sunlight spectrum, materials with bandgap energies in the near-infrared region (0.8–1.1 μm, approximately 1.1–1.6 eV) show optimal efficiency. For this reason, Si (Eg = 1.1 eV) and GaAs (Eg = 1.4 eV), which have absorption edge wavelengths suitable for the solar spectrum, are used in solar cells.

Considering the above situation, as shown in Figure 5.4, the efficiency can be improved by irradiating monochromatic light with a wavelength close to the bandgap (absorption edge) of the solar cell. This is because it produces a small amount of thermal energy but absorbs enough light; in the case of OWPTs, the monochromatic light can be selected and the wavelength of this light can be appropriately chosen. In fact, the laser is a monochromatic light with a spectral width of at most a few nm (ΔE < 1 meV).

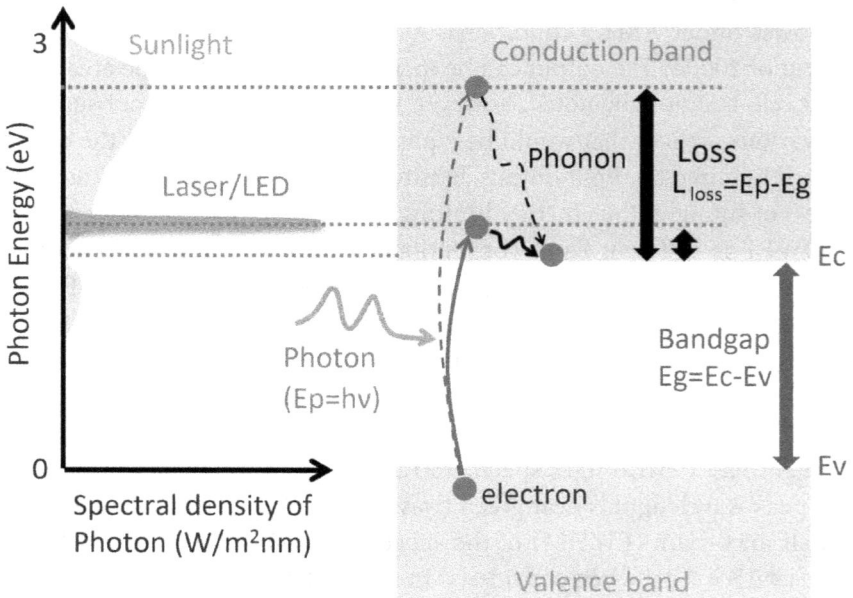

FIGURE 5.4   Monochromatic light irradiation of solar cells.

In addition, the efficiency of solar cells typically increases with the density of the light intensity. Unlike sunlight, where the light intensity is fixed, in OWPTs the light intensity density can be designed. As a result, efficiencies of 40–70% are expected when monochromatic light of tens to hundreds of times the intensity of sunlight is irradiated, as discussed below. Compared to sunlight power generation with an efficiency of only about 20%, the same solar cell devices in OWPTs can operate at very high efficiencies.

5.2.3 Basic Physics and Characteristics of Solar Cells

5.2.3.1 Electrical Characteristics of Solar Cells

This section describes in detail the basic operation of a solar cell. In this chapter, we refer to them as "solar cells" even though they do not irradiate sunlight.

First, the basic and important functions of solar cells are confirmed. Electricity is generated by light irradiation. Electricity is expressed as the product of voltage and current. Therefore, solar cells must be able to extract both voltage and current. Note that although there are various photo-receiving devices, they basically only output either voltage or current in their basic operating state and cannot be substituted for solar cells.

A solar cell absorbs light incident on a semiconductor pn-junction to generate charge carriers (electrons and holes), and then extracts the generated charge carriers as an electric current along with a voltage related to the bandgap to an external circuit for use as electrical power. This function is similar to that of a photodiode (PD) using a pn-junction, so the basic principle is explained including the differences from a PD. A schematic diagram of the current–voltage characteristic (I–V curve) of a pn-junction is shown in Figure 5.5.

A reverse bias voltage is applied to the pn-junction in the PD, and a current corresponding to the absorbed light is output. Although current can be extracted without reverse bias, reverse bias is applied to properly extract the carriers (current). The extracted current is converted to a voltage by an electrical road resistance, and voltage detection is often performed. Solar cells, on the other hand, do not require an external voltage to extract power. In other words, in addition to the current corresponding to the amount of light, the voltage generated by the solar cell itself is also important.

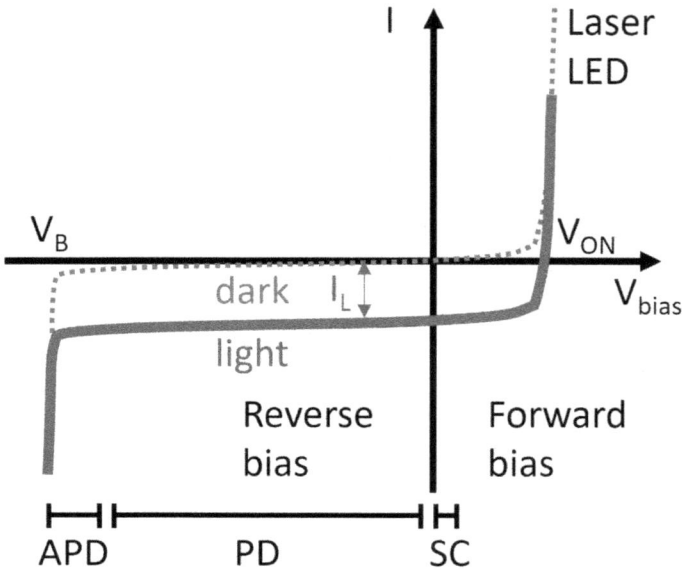

FIGURE 5.5   Schematic diagram of I-V curve of pn-junction.

Although solar cells and PDs have different operating conditions, the same devices can be used in principle because they have the same pn-junction. However, in the case of PDs, in addition to the amount of current, which is closely related to sensitivity and efficiency, the speed of response to signals is also important. In a typical vertical-incidence PD, efficiency alone cannot be emphasized because the two are in a trade-off relationship. On the other hand, solar cells do not require speed of response to signals (light). Therefore, solar cells can be optimized for efficiency. It should be noted that due to the large device size of solar cells, the speed of response to signals is basically very slow. In OWPT systems, communication using light for power transmission is sometimes considered, however when solar cells are also used as optical signal receivers, high-speed communication is problematic.

The APD in the figure is an avalanche photodiode; an APD is a PD whose bias voltage is set near the breakdown voltage to obtain current gain. In addition, the semiconductor lasers and LEDs in the figure also use pn-junctions, but the applied voltage is the forward voltage.

In the case of solar cells, when light incidents on the pn-junction and is absorbed, a photocurrent flows that is used as the output current. The photocurrent is a negative current in the opposite direction of the current

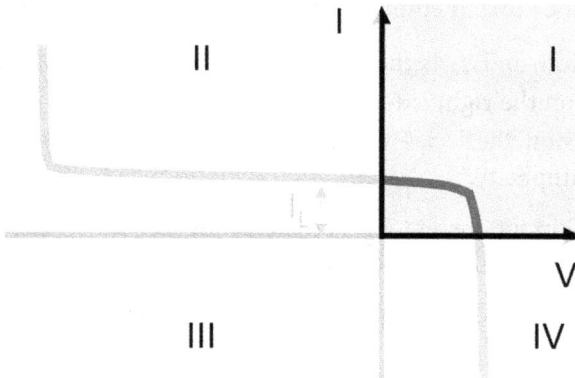

FIGURE 5.6 The first quadrant of the I-V characteristics of a solar cell.

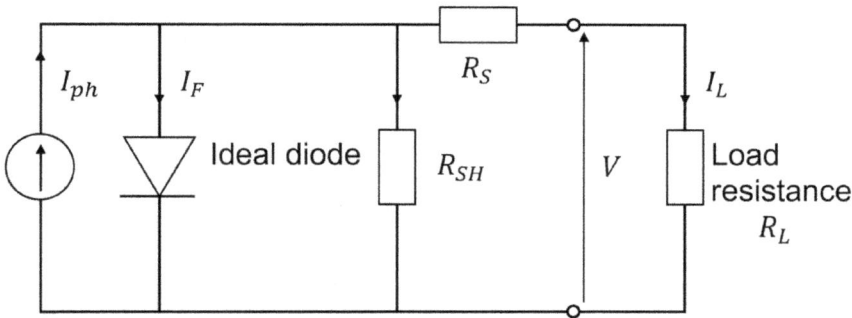

FIGURE 5.7 Equivalent circuit of a solar cell when connected to a load.

of I–V characteristic of the diode when a forward voltage is applied. In practice, light irradiation will cause the curve of the characteristic to shift downward, as shown in the figure. For this reason, the I–V characteristic of a solar cell is focused only on the fourth quadrant of the I–V characteristic of a typical diode, i.e., the first quadrant of the I–V characteristic drawn upside-down as shown in Figure 5.6.

Figure 5.7 shows the equivalent circuit of a solar cell connected to a load. Based on this circuit, the following equations are given.

$$I_L = I_{ph} - I_0 \cdot \left(exp\left(\frac{eV}{nkT} \right) - 1 \right) \tag{5.1}$$

$$V = I_L \cdot R_L \tag{5.2}$$

where I_L is the current entering the load, I_{ph} is the current generated by light absorption, and I_0 is the reverse saturation current of the diode. The second term on the right side is the current flowing in the ideal diode, I_F. In this discussion, the series resistance R_S and the shunt resistance R_{SH} are ignored for simplicity.

The short-circuit current I_{sc} is the current that shorts the terminals of the solar cell and is ideally equal to the current based on the incident light power because of $V = 0$, as shown in the following equation.

$$I_{sc} = I_{ph} \tag{5.3}$$

The open-circuit voltage V_{oc} is the voltage measured with the solar cell terminals open, which is close to the diode turn-on voltage and is given by the following equation

$$V_{oc} = \frac{nkT}{e}\ln\left(\frac{I_{ph}}{I_0}+1\right) \approx \frac{kT}{e}\ln\left(\frac{I_{ph}}{I_0}\right) \tag{5.4}$$

Normally, this open-circuit voltage will be close to the bandgap energy of the pn-junction, however, in detail, the open-circuit voltage will be slightly smaller than the bandgap energy.

When solar cells are connected to a load, they operate at the operating point determined by the load characteristics (impedance or resistance). In other words, the operating point is somewhere in the point of the middle range of the I–V curve from the short-circuit current to the open-circuit voltage, as shown in Figure 5.8. At the open-circuit voltage point, the current is zero, and at the short-circuit current point, the voltage is zero, so the output power is zero. If the circuit can be operated at the optimum operating point where maximum power can be extracted, it will operate at high efficiency. The optimum operating point is near the shoulder of the I–V curve, and this optimum operating point has a current slightly less than the short-circuit current and a voltage slightly less than the open-circuit voltage. Even though the load characteristics are defined on the load circuit, the maximum operating point varies depending on the amount of

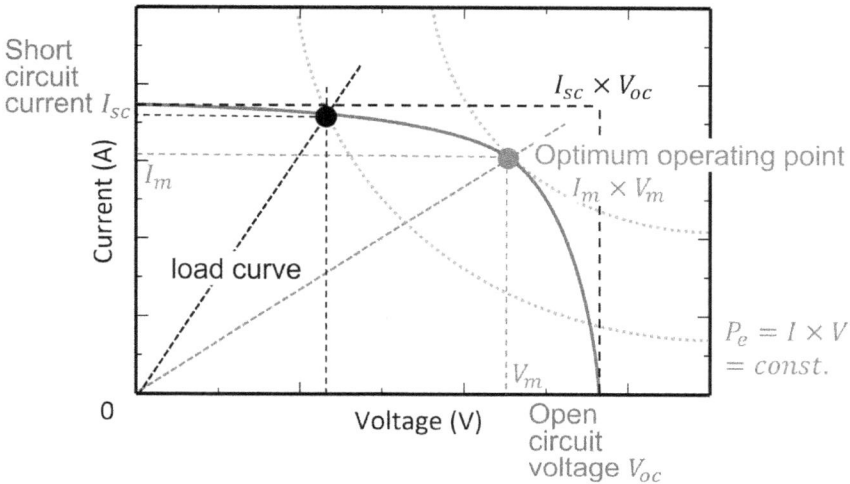

FIGURE 5.8   Operating points in the I-V curve of a solar cell according to the load characteristics.

light irradiated and the required load resistance is changed. For this reason, the maximum power point tracking (MPPT) function (circuit) is used in photovoltaic power generation.

The fill factor (FF) is also an important characteristic of the solar cell, which is the ratio of the area of the rectangle created by I_m and V_m at the optimum operating point to the area of the rectangle created by I_{sc} and V_{oc} as shown in the figure. The area of the rectangle corresponds to the power. The closer FF is to 1, the closer the I–V characteristic curve approaches the edge of the rectangle bounded by the open-circuit voltage and short-circuit current. This FF is mainly influenced by the characteristics of the resistors and diodes inside the solar cell. It is generally around 0.7 (0.6–0.8).

5.2.3.2 Efficiency of Solar Cells

Figure 5.9 shows the relationship between the energy irradiated as light and the amount of energy that can be extracted as power. It is basically an I–V characteristic similar to Figure 5.8, where the x-axis is the voltage and shows the photon energy, the bandgap energy, and the open-circuit voltage. Here, the photon energy $\hbar\omega$ and the bandgap energy Eg have units of eV, but are converted to voltage by dividing by the electron charge e. The

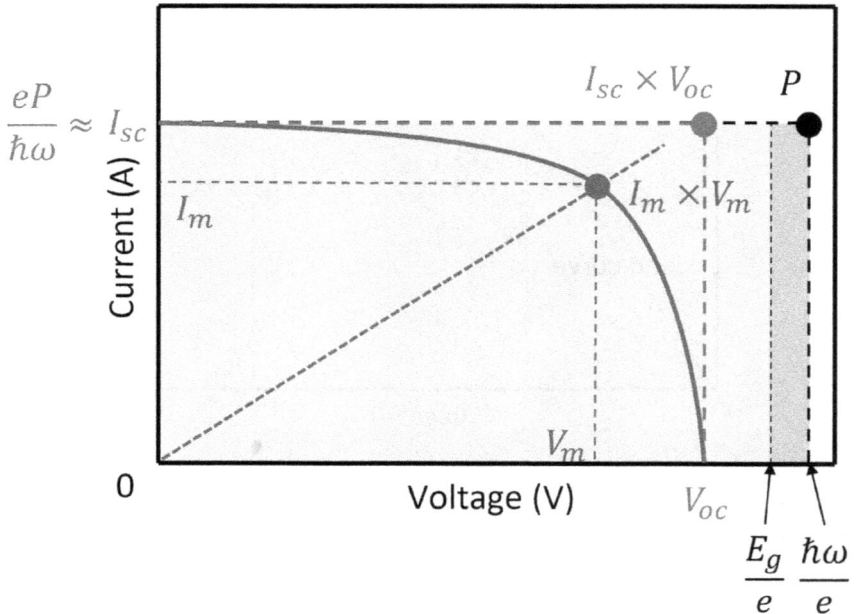

FIGURE 5.9   Relationship between electrical characteristics, bandgap, and irradiated light power.

y-axis shows the current, with the number of photons and the short-circuit current. The number of incident photons per unit time is $P/\hbar\omega$, which can be converted to current by multiplying it by the electron charge.

Under ideal load conditions, the output power is calculated as $I_m \times V_m$. The irradiated light power P is calculated by multiplying the number of photons (per unit time) by the photon energy. Therefore, the power conversion efficiency η of a solar cell is expressed as

$$\eta = I_m \times V_m / P \tag{5.5}$$

It can also be expressed as

$$\eta = I_{sc} \times V_{oc} \times FF / P \tag{5.6}$$

The actual short-circuit current I_{sc} is smaller than the current corresponding to the number of incident photons N_{ph}. The ratio I_{sc}/eN_{ph}

is called the current efficiency and is commonly referred to as the external quantum efficiency (EQE). The open-circuit voltage V_{oc} is less than the voltage corresponding to the incident photon energy $\hbar\omega$. This ratio $V_{oc}/(\hbar\omega/e)$ is called the voltage efficiency. Considering the relationship between the incident light power P and the number of incident photons N_{ph}, denoted by $N_{ph} = P/\hbar\omega$, the power conversion efficiency η can also be expressed as follows,

$$\eta = \left(I_{sc}/eN_{ph}\right) \times \left(V_{oc}/\left(\hbar\omega/e\right)\right) \times FF \qquad (5.7)$$

In other words, the power conversion efficiency η is expressed as the product of the current efficiency (EQE), voltage efficiency, and FF.

Figure 5.10 shows the calculation results of the open-circuit voltage and bandgap energy, as well as the voltage efficiency and bandgap energy, based on the simple theoretical model of the solar cell. The analysis method used is based on the well-known Shockley–Queisser equation [4], taking into account the effective mass approximation of semiconductor band structures as discussed by K. Iga and G. Hatakoshi [5].

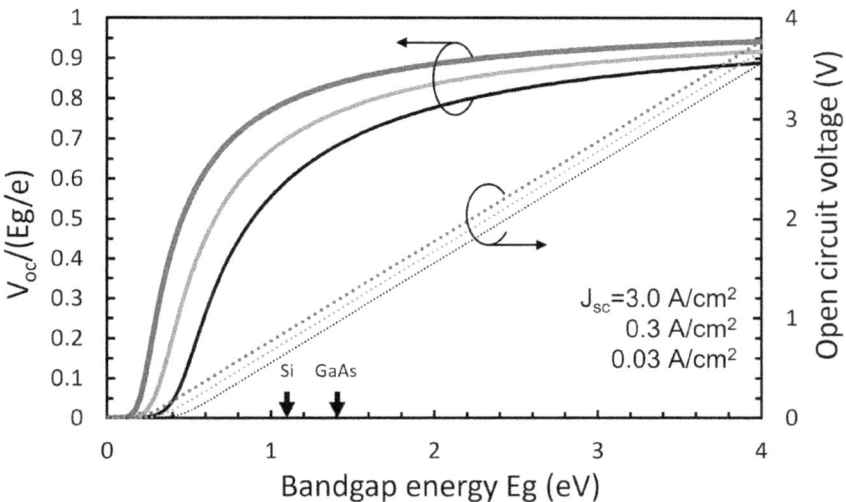

FIGURE 5.10 Relationship between voltage efficiency, open circuit voltage, and bandgap energy.

In this figure, the voltage efficiency is calculated using the open-circuit voltage and the bandgap energy for some different short-circuit currents. Although there are many ways to compare sunlight and monochromatic light, such as the same irradiation power density, the same short-circuit current, and the same output power conditions, in the figure, the short-circuit current of 0.03 A/cm² is set to be approximately equivalent to the current under sunlight irradiation.

The materials with bandgap energy shown on the x-axis are set as hypothetical materials considering some of the existing III–V compound semiconductors. As a result, the open-circuit voltage is almost proportional to the bandgap energy, and it is found that the open-circuit voltage is about 0.2–0.5 V lower than the voltage corresponding to the bandgap energy. This important finding leads to the method of improving voltage efficiency by using wide bandgap materials. On the other hand, the voltage efficiency decreases drastically for narrow bandgap materials. These findings are important factors in determining the wavelength band to be used for OWPT.

The ideal efficiency of each bandgap material can be obtained by multiplying the voltage efficiency by an *FF* of about 0.7, assuming a current efficiency of 1. As a result, assuming a monochromatic light intensity equivalent to sunlight (I_{sc} = 0.03 A/cm²), an efficiency of about 40% is expected for Si and about 50% for GaAs. This is about twice the efficiencies of 20% and 25%, for sunlight irradiation conditions, respectively, for the same solar cells under sunlight irradiation conditions.

In addition, the efficiency of solar cells tends to increase as the intensity density of light increases. Considering an intensity 100 times that of sunlight, an efficiency of 50–60% or more can be expected even for Si and GaAs. As the light intensity continues to increase, the efficiency decreases due to temperature rise.

In addition, as discussed above, if solar cells with shorter wavelengths are prepared by preparing light sources for corresponding wavelengths, high efficiencies can be expected in green, blue, and ultraviolet. At present, there is little research being done on solar cells in these shorter wavelength bands. This is because the importance of these short-wavelength solar cells to sunlight is reduced. From the perspective of the OWPT, the research and development of such short wavelength solar cells is extremely attractive.

The efficiency of solar cells with respect to monochromatic light has been reported as follows. For the near-infrared region, efficiencies of over 40% have been reported for Si solar cells with the wavelength band of 900 nm

[6]. In the case of GaAs solar cells with a wavelength band of 800 nm, efficiencies of 68.9% (@9.6 W/cm²) have been reported for a single-junction structure [7]. In addition, an efficiency of 66% has been reported for a multi-junction GaAs structure [8]. Even in the long wavelength range of 1470 nm, where the efficiency is inherently prone to decrease, an efficiency of about 50% has been reported using a multi-junction InGaAs/InP system [9]. These high-efficiency solar cells are promising for high-performance OWPT systems. On the other hand, GaN-based materials have recently been investigated for short-wavelength OWPT, with a reported efficiency of 42.7% [10].

An interesting configuration of solar cells is discussed. For solar cells used in sunlight, a multi-junction structure is sometimes used for high efficiency. In a multi-junction structure, several materials with different bandgaps are stacked to cover a wide range of sunlight wavelengths. Multi-junctions using two to six different materials have been investigated, and it has been reported that a maximum of 47.1% can be achieved under concentrated sunlight with a six-junction structure [11]. However, in the case of OWPTs, such a conventional multi-junction structure may be ineffective or even disadvantageous because multiple wavelengths of light need to be prepared with appropriate intensity ratios for each wavelength. On the other hand, multi-junctions using the same absorbing material have been developed for OWPTs as discussed above [8,9]. Such structures have the advantage of higher output voltage and higher efficiency. Some such multi-junction solar cell devices have already been commercialized.

5.2.4 Features of Light Sources

Light from sunlight, flames, candles, and gas lamps, as well as incandescent, fluorescent, and discharge lamps, has been used primarily for lighting for many years. Today, there are also light-emitting diodes (LEDs), semiconductor lasers, and other types of lasers that are also used as light sources.

For application in OWPTs, the emitted light must be easy to use as a beam, must be able to produce a high output corresponding to the required electrical power, and must have a sufficiently high efficiency. Based on these basic requirements, LEDs and semiconductor lasers are effective in terms of high efficiency, high output power, small solid-state devices, and direct electrical drive. Among lasers other than semiconductor lasers, fiber lasers are also considered to be effective as one of the light sources for OWPT due to their relatively small size, high output power, and excellent light beam quality. Although there are various characteristics and features of light sources, in the application to OWPT, light output power, efficiency, and light beam characteristics are mainly discussed.

5.2.4.1 High-power Laser Light Sources

High-power light sources are mainly used for LEDs for lighting and displays, and semiconductor lasers for processing, heat treatment, optical pumping of other lasers, next-generation lighting, next-generation displays, and LiDAR (laser imaging detection and ranging), and so on. R&D activities have been intensified in line with the expansion of these applications and markets, and improvements in characteristics, i.e., higher output power, are still in progress.

The basic design for increasing light output is to increase the size of the device. In general, there is always a loss in a device, and the heat generated by the loss saturates the maximum output of the device. This limits the ability to achieve high output power with small devices. For LEDs, a single semiconductor chip of a few millimeters square provides several watts of light output; for semiconductor lasers, a single semiconductor chip of a few millimeters square provides several watts to several tens of watts of light output. LEDs use several of these chips to achieve high brightness lighting, while semiconductor lasers use arrays, chip stacking modules as shown in Figure 5.11, and fiber-coupled modules as shown in Figure 5.12 to achieve light outputs of several hundred watts to several kW. Such a fiber-coupled module is called a fiber-coupled laser or a direct diode laser (DDL). Increasing the efficiency of the elements is extremely important in reducing the number and size of devices to achieve such high output power.

FIGURE 5.11   Types of high-power semiconductor lasers. Single, bar (1D-array), stacked.

Semiconductor laser

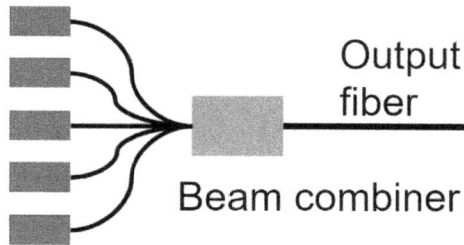

FIGURE 5.12   Fiber-coupled high-power lasers.

In addition, by using several of these high-output modules, light output in the 100 kW class or higher can be achieved in proportion to the number of modules. Thus, the light output itself is at a stage where it can cover almost the entire electrical power range required for OWPT systems, from the mW class to the multiple 10 kW class.

In the case of lasers, such high-power lasers have been put to practical use mainly in the near-infrared range of wavelengths from 800 to 1000 nm, and high-performance lasers of more than 10 kW are available. In the visible range, especially in the blue light range, several watts of output power per chip have been commercialized. Similarly, high output power has been developed by fiber coupling, and blue light sources of several kilowatts have been commercialized.

In addition to edge-emitting lasers, as shown in Figure 5.11, which emit light from the edge of a semiconductor substrate, vertical-cavity surface-emitting lasers (VCSELs) [12], which emit light from the substrate surface, are among the semiconductor lasers that have recently been applied to several new photonic applications and have created a large market [13].

VCSELs are realized as ultra-compact laser devices with emission area diameters of 3–15 μm. They are used as ultra-low-power consumption lasers with light output in the mW class or less, one to two orders of magnitude smaller than edge-emitting lasers. Due to their low power consumption, VCSELs are used in sensor applications such as laser mice, camera autofocus, and proximity sensors, as well as in short-haul fiber communications. Modules for communications that are independently driven by multiple arrays of individual devices and high-speed, high-resolution laser printers are in practical use.

On the other hand, by fabricating a large number of 2D array VCSELs on a single semiconductor chip and driving them all at once, it is possible

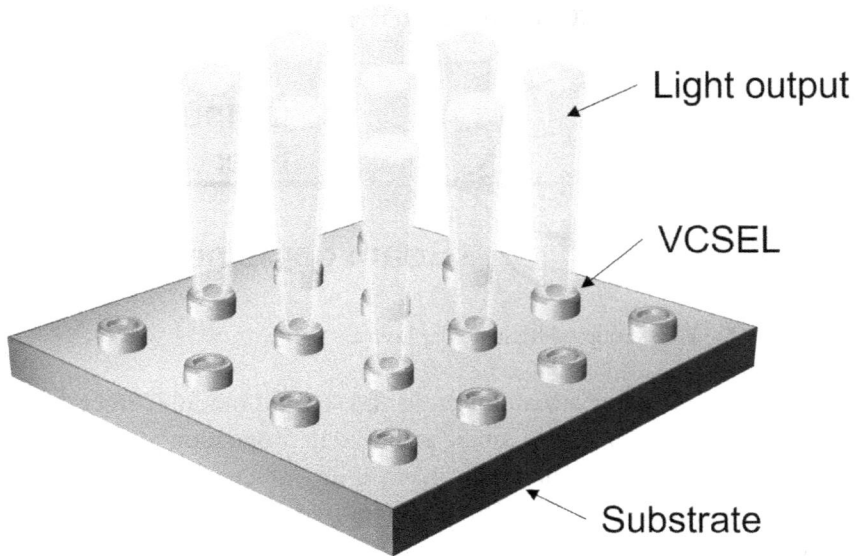

FIGURE 5.13   Schematic structure of a two-dimensional array VCSEL.

to achieve an effectively large area and high output power, as shown in Figure 5.13. Such high-power 2-D arrays are now beginning to be used in large quantities as infrared illuminators for sensing applications, such as face recognition systems and flash-type LiDARs. In other words, VCSELs can be used as high-power light sources through 2-D array configuration.

In the case of VCSELs, the main wavelengths in practical use in 2D arrays are in the near-infrared range from 0.8 to 1.1 μm, and several tens of W or higher output powers have been realized and commercialized. Long-wavelength devices such as 1.3–1.5 μm are also in practical use, but the long-wavelength band is mostly used as a single device. On the other hand, in the short-wavelength band, red-color devices have been commercialized, and GaN-based VCSELs for blue-violet are still in the research and development stage. However, in recent years, the performance of GaN-based VCSELs has improved remarkably [14]. The GaN material system has excellent temperature characteristics, so once a certain performance is achieved in a single device, it will be possible to increase the number of arrays to achieve higher output power in a short period of time.

In addition to VCSELs, photonic crystal lasers that can also emit light from the surface of a semiconductor substrate have been investigated [15]. These photonic crystal surface-emitting lasers (PCSELs) are capable of

Semiconductor laser
for excitation

Fiber amplifer

Mirror

Mirror

Beam combiner

Output
fiber

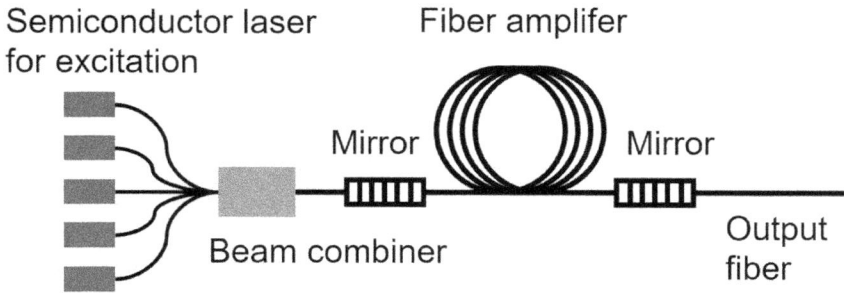

FIGURE 5.14   Schematic configuration of a fiber laser.

operating as a single laser, regardless of the area of the device, and thus have the advantage of emitting a high-quality beam. In recent years, PCSELs have also become important as high-power, high beam quality lasers.

In addition to the fiber-coupled laser shown in Figure 5.12, which directly couples the light output of high-power semiconductor lasers, there is a fiber laser that uses the optical fiber itself as a laser cavity, as shown in Figure 5.14. In this fiber laser, light from the semiconductor laser is injected into the optical fiber, and this excitation light is used as an energy source to generate new laser light in the optical fiber. Fiber lasers can increase the output power using the volume size (length) of the fiber itself, generate short pulses using nonlinearities, and broaden the spectrum. In OWPT, continuous power transmission is typically considered to be important. High-power CW (continuous wave) fiber laser systems in the 10 kW class are already commercially available and used for processing.

5.2.4.2 Efficiency of Lasers

To understand the laser efficiency that follows, a simple model of the I–L–V (current–light-output–voltage) characteristic of a semiconductor laser is shown in Figure 5.15. Since a light-emitting device is essentially a current-driven device, the x-axis represents the current. The I–L characteristic provides a linear light output for a current above a threshold value. In addition, because the diode uses a pn-junction, current flows at or above the diode turn-on voltage in the I–V characteristic. Note that because of the electrical resistance component, the voltage gradually increases as the current increases. Efficiency is obtained from the relationship between input power based on the product of current and voltage and light output.

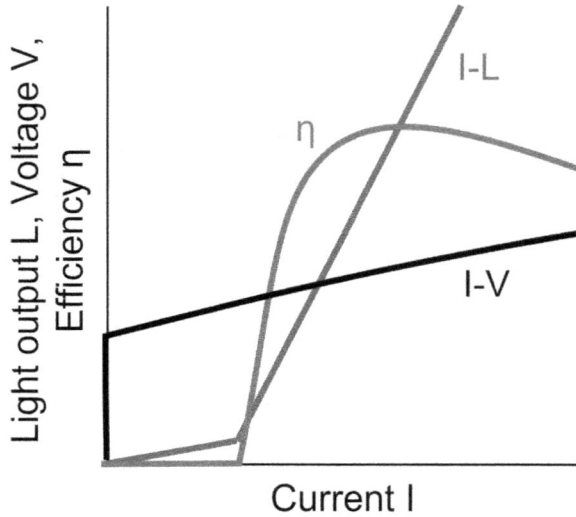

FIGURE 5.15   Schematic I-L-V characteristics of semiconductor lasers.

A simple operating model can be expressed in the following equations.

$$P = SE\left(I - I_{th}\right) \tag{5.8}$$

$$V = RI + V_0 \tag{5.9}$$

$$SE = \frac{\hbar\omega}{e} \cdot \eta_d = V_0 \cdot \eta_d \tag{5.10}$$

$$\eta = \frac{P}{VI} \tag{5.11}$$

where P is the light output, SE is the slope efficiency of the light output expressed in units of W/A, I is the drive current, I_{th} is the threshold current of the laser, V is the applied voltage, V_0 is the pn turn-on voltage, R is the resistance, [ω is the photon energy, e is the electronic charge, η_d is the external differential quantum efficiency, η is the electro-light conversion efficiency (efficiency), and VI is the injection power. As shown in Figure 5.15, the efficiency peaks relatively close to the threshold current, and increasing the light output power beyond this point does not increase the efficiency.

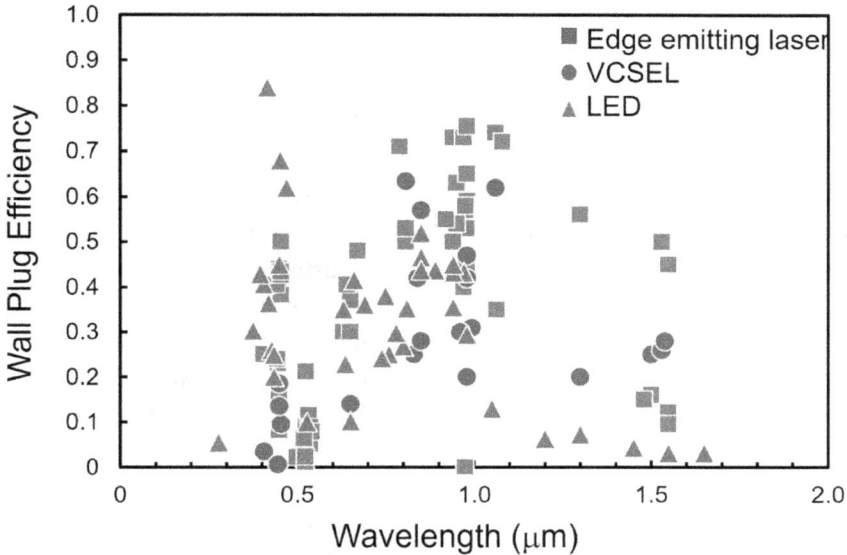

FIGURE 5.16   Efficiency of LED and semiconductor lasers.

Examples of actual electrical light conversion efficiencies (PCE: power conversion efficiency; WPE: wall plug efficiency) of edge-emitting lasers, VCSELs, and LEDs are shown for wavelengths in Figure 5.16. For lasers, the values are mainly those reported in journal articles. For LEDs, commercial devices were referenced in addition to some values reported in journal articles.

As can be seen from the efficiency examples shown in the graph, edge-emitting lasers have been reported to have efficiencies exceeding 40% over a wide wavelength range from ultraviolet (0.4 μm) to infrared (1.6 μm). In particular, there are several reports of efficiencies exceeding 70% in the near-infrared band from 0.8 to 1.1 μm [16–18]. In the short wavelength band of 450 nm, an efficiency of 50% has been reported. However, in the green and yellow wavelengths of 0.5–0.6 μm, the efficiency is very low due to the difficulty in obtaining high-quality materials and material properties.

For applications such as processing, high output power as a single device is important from the point of view of light beam controllability. In the design of semiconductor lasers, there is a difference between designs that increase output power as a single device and designs that increase maximum efficiency. In other words, conventional devices have been designed to prioritize high output power and then increase efficiency. In some applications of OWPTs, it is important to prioritize efficiency, and there are designs for this purpose. If efficiency is the only priority, it is in principle possible to achieve an efficiency of 80% or more with a short cavity length. Such a design may not provide sufficient light output, however, in OWPTs where the light beam characteristics are less constrained by the relatively short power transmission distance such as in indoor applications, the required light output can be achieved by using multiple devices.

There are several reports of VCSEL efficiencies of 60% or higher [20,21], and an increasing number of commercial devices are achieving 40%. However, this efficiency value itself is about 10 points lower than the efficiency of edge-emitting lasers. This may be due to the vertical emission structure of VCSELs.

For LEDs, the efficiency seems to be slightly lower than that of semiconductor lasers, but for blue, it is higher than that of semiconductor lasers. The efficiency of LEDs is comparable to that of semiconductor lasers for several applications.

As for the efficiency of fiber lasers, the light emitted from a semiconductor laser is introduced into a fiber as excitation light for lasing, which requires a two-step energy conversion that reduces the overall efficiency. However, since the conversion efficiency of the excitation light into fiber laser light is relatively high, and the efficiency of the semiconductor laser for the excitation light is also high, an overall efficiency of 30–50% can be achieved. Considering the use of advantageous characteristics of fiber lasers, this can be considered an acceptable efficiency.

5.2.4.3 Beam Characteristics and Long-distance Transmission

Some applications other than OWPT may require advanced control of light beam characteristics, such as focusing light to a spot as small as a millimeter or less, or emitting a light beam with a small spread over a long distance. In such cases, it is important to achieve high output power from a single device or module capable of emitting a high-quality beam. In OWPT, a high-quality beam is required for long-distance transmission, however, the light can be irradiated over a relatively large area. For relatively short distance transmission, the required beam quality is reduced.

Light beams emitted into free space, even laser beams, will diverge due to diffraction. Therefore, it is necessary to use optics to focus the light beam to an appropriate size, especially for long-distance transmission of OWPT. In addition, the shape of the light beam must be controlled to transmit the light to the light receiver. At this point, the light beam characteristics of the light source itself limit the focusing performance and controllability of the irradiated light beam even when the laser is used.

Light beam characteristics or beam quality here refer to the inherent spatial distribution of light as an electromagnetic wave. A light beam with the same phase plane and excellent spatial coherence has high beam quality, and a single-peak Gaussian beam (Gaussian function shape) is generally considered to be ideal for a light beam radiated in free space. In the case of an ideal Gaussian beam, although there is still diffraction, the beam size can be maintained at a long transmission distance by using the collimation condition.

The light emission pattern of LEDs is also bright at the center and fades further from the center, but it is quite different from the Gaussian beam described above. In other words, a beam with excellent spatial coherence cannot be obtained with an LED, but only with a laser. In addition, the beam quality deteriorates when lasers produce multiple transverse-mode beams simultaneously. Therefore, a single transverse mode laser is required to achieve high beam quality. A single transverse mode light beam can be focused to a very small spot down to about one wavelength range by using an appropriate lens. The high-efficiency focusing characteristics of light with poor spatial coherence, such as LEDs and multiple transverse mode lasers, are limited to the size of the emission area.

Here, the beam size in long-distance transmission is analyzed when an ideal Gaussian beam is used. The Gaussian beam is shown to be

$$f_1(x', y', 0) = E_0 \exp\left(-\frac{1}{2} \cdot \frac{x'^2 + y'^2}{s^2}\right) \tag{5.12}$$

where f_1 is the electric field of the light, s is the output spot size and $1/e$ of the peak intensity. The light intensity distribution is proportional to the square of the electric field. The spot size can be controlled by the lens system of the beam expander. After propagating a distance z, the beam spot size w is

$$w = s\sqrt{1 + \left(\frac{z}{ks^2}\right)^2} \cong \frac{z}{2\pi} \cdot \frac{\lambda}{s} \tag{5.13}$$

where k is the wavenumber.

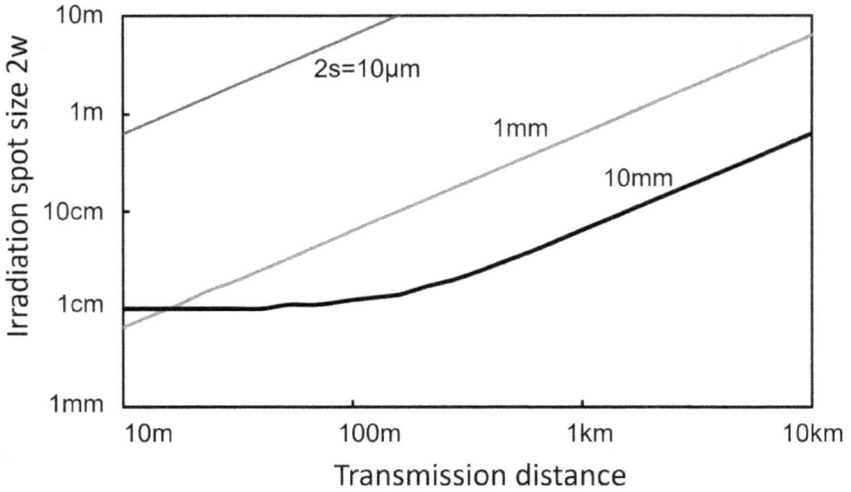

FIGURE 5.17 Dependence of beam size after long-distance propagation on output beam size.

Figure 5.17 shows the beam spot size after different transmission distances for different output beam spot sizes. For an output beam spot size of 10 μm, which corresponds to the beam size of a semiconductor laser or a single-mode fiber laser, the beam spot after propagation at a distance of 100 m increases to the 10 m class.

On the other hand, if the output beam spot size is increased to 10 mm using a beam expander, the transmitted beam spot size after 100 m is almost 1 cm, and the beam spot size after 1 km is still in the 10 cm class. In this analysis, only the diffraction of the output beam was considered, but when focusing optics are used, the beam size becomes even smaller. On the other hand, since the beam energy extends beyond the spot size region, a solar cell two to three times larger than the result in the figure would be required to collect enough light. These results indicate that, under ideal conditions, a transmission distance of up to 1–10 km is feasible for the OWPT, depending on the size of the light receiver.

Under conditions that do not require long-distance transmission and allow light beams of a certain size, the requirements for the quality of the light beam are low. For such applications, light sources with poor beam quality such as LEDs and VCSEL arrays can be used.

For a light beam to be properly transmitted over long distances, the laser should have a single transverse mode and an M^2 value close to 1, which indicates the quality of the beam. The M^2 value indicates how many times the beam size can be reduced down to the diffraction limit.

$$M^2 = w\theta\frac{\pi}{\lambda}$$ (5.14)

where λ is the wavelength. The spot size w is defined as the radius of $1/e^2$ of the peak intensity for an ideal Gaussian beam or the second moment of the light intensity for a typical beam.

The beam parameter product BPP, which is the product of the radius of the beam waist and the half-width of the beam spread angle

$$BPP = w\theta = M^2\frac{\lambda}{\pi}$$ (5.15)

is sometimes used. For a typical edge-emitting laser, BPP is about 4–5 mm/mrad in the high light output power range. If the laser beam is of high quality with a single transverse mode, beam divergence can be easily suppressed by using a simple beam expander. As discussed earlier, it is important that the light beam is a high-quality single transverse mode beam in order to use a collimated light beam that does not change much in beam size over long distances.

Edge-emitting lasers are usually designed to have a single transverse mode in the thickness direction (fast axis) of the semiconductor substrate. On the other hand, the transverse mode in the stripe width direction (slow axis) becomes a single transverse mode when the stripe width is about a few micrometers or less, which is about the wavelength size. However, the transverse mode becomes multiple transverse modes when the stripe width is larger than this size. In the case of high-power lasers, since the emitting layer volume is small for a narrow stripe width, high power is generally achieved with a wide stripe width, resulting in a multi-transverse mode. In this case, the beam controllability is severely degraded due to the multi-transverse mode.

Arrayed VCSELs operate similarly to a combination of many transverse modes and their beam quality is very poor. Due to their spatial incoherence, the beam quality of arrayed light sources is almost equivalent to that of LEDs in their emission area, except for the beam divergence angle.

Fiber lasers can be designed as single-mode or multi-mode depending on their configuration and the fiber core design. Since the laser light is generated based on the incident excitation light, it is possible to generate the desired mode independent of the mode characteristics of the incident excitation light. On the other hand, fiber-coupled lasers (direct diode lasers) combine light from multiple semiconductor lasers. When multiple single transverse mode semiconductor lasers are optically coupled to a multi-mode fiber, the output from the end of the fiber is multi-transverse mode. If the lasers have different wavelengths and are orthogonally polarized, they can be coupled as a single transverse mode with appropriate optics.

As mentioned above, the better the quality of the light beam, the easier it is to focus and control, which is advantageous for longer transmission distances. On the other hand, to produce high-power light as a high-quality single transverse mode beam, the beam size must be as small as the wavelength. The need to reduce the beam size reduces the volume of the device and severely limits the light output. Long structures, such as fiber lasers, are advantageous or essential for high-quality beams with high output powers of 10 W or more.

Finally, it is considered how LEDs can be used for OWPT. LEDs with appropriate chip size have output power close to that of semiconductor lasers and efficiency close to that of semiconductor lasers, and the efficiency of solar cells is comparable to that of semiconductor lasers, although the spectral width of monochromatic LEDs is wider than that of lasers. However, LEDs have poor beam characteristics, mainly due to their principle of emitting light in whole directions, and require relatively large lenses to achieve narrow beam angles. Although the application for long-distance transmission will be limited, it is possible to use LEDs in OWPTs, under conditions where these beam characteristics and the lens system are not a problem.

5.2.5 Efficiency of Optical Wireless Power Transmission

5.2.5.1 Expected Efficiency as Upper Limit Performance

The previous sections describe the respective efficiencies of light sources and solar cells. This section discusses the power transmission efficiency of OWPT based on these characteristics.

The detailed components of the power transmission efficiency include the conversion efficiency from the electricity power source to the light beam, losses due to beam shaping and focusing, transmission losses in the

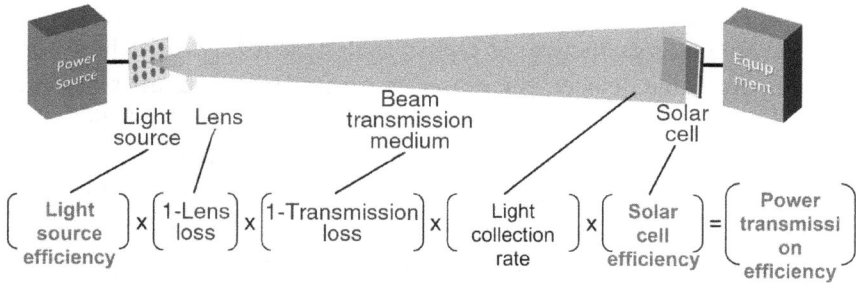

FIGURE 5.18 Components involved in the power transmission efficiency of the OWPT.

TABLE 5.1 Expected (upper limit) Power Transmission Efficiency of OWPT

		Efficiency of laser	Efficiency of solar cell	Power transmission efficiency
Current	Typical of commercial use	40%	40%	16%
	Maximum reported value	75%	69%	52%
Future		85%	85%	72%

transmission medium, the collection rate, which is the ratio of the beam entering the beam receiver at the transmitted distance, and the conversion efficiency from light to electricity power, as shown in Figure 5.18. As mentioned here, there are also losses in the transmission medium and devices other than the light source and solar cell, such as lenses, but the effects of these related parts vary greatly depending on the system configuration and application conditions. Therefore, only the effects of the light source and solar cell are considered as the basic elements of the power transmission efficiency. This basic power transmission efficiency is calculated as the product of the electrical-to-optical conversion efficiency of the light source and the optical-to-electrical conversion efficiency of the solar cell, and can be considered as the upper limit of the power transmission efficiency in various applications and configurations.

Table 5.1 shows the expected values of power transmission efficiency. Based on the characteristics of the current typical commercial laser and solar cell devices and the maximum reported characteristics, the product of their efficiencies was taken as the power transmission efficiency. Commercial solar cell devices are assumed to be typical silicon or GaAs solar cells. As mentioned above, the wavelength of the laser and the wavelength (bandgap) of the solar cell must be properly matched. Fortunately, the efficiency of near-infrared lasers effective for typical solar cells is high. For these commercially available devices, the power transmission efficiency is expected to be around 10–20%. On the other hand, the upper limit of the power transmission efficiency is expected to be 40–50% for the case assuming almost the maximum efficiency of the reported devices. This performance will be practically available in the near future.

On the other hand, the table also shows expected values for other future devices. The efficiency of solar cells is assumed to be high at short wavelength bands. In this case, the short-wavelength light source should also be prepared as high efficiency at the same time. In this prospective case, the upper limit of power transmission efficiency is expected to reach about 70%.

As described above, it is unlikely that OWPT based on a semiconductor light source and semiconductor solar cells will be able to achieve a power transmission efficiency of more than 90% like induction-coupled WPT, even in the future. However, it should be noted that OWPT offers values that cannot be realized by other methods because of its wireless power transmission function, small change in power transmission efficiency over long distances, and other features. However, it is also important to continue efforts to improve the efficiency of devices in order to increase the value.

Although not explained in detail here, the efficiency of both semiconductor light sources and solar cells increases at lower temperatures. Therefore, depending on the environment in which they are used, the efficiency may be higher than the efficiencies listed above.

5.2.5.2 Effect of Transmission Distance on Efficiency

This section briefly describes the transmission losses that occur over a transmission distance. Specifically, transmission losses due to absorption and scattering in the transmission medium are considered. These losses limit the transmission distance.

Absorption loss is caused by molecules in the transmission medium, i.e., air. Visible light has almost negligible absorption, but ultraviolet light is absorbed at certain wavelengths by nitrogen and oxygen molecules, and infrared light is absorbed by water vapor, carbon dioxide, methane, and ozone. The wavelength band with little atmospheric absorption is called the atmospheric window wavelength.

In addition to absorption loss, scatter loss also affects transmission loss. Rayleigh scattering is a phenomenon in which light is scattered by particles that are less than one-tenth the size of the light wavelength. The sky appears blue in daylight because of the wavelength dependence of Rayleigh scattering in the visible range by molecules in the air. Another phenomenon is Mie scattering, in which light is scattered by particles that are about the same size as the wavelength of light. Clouds appear white because Mie scattering occurs almost uniformly throughout the visible range due to the size of the water particles that make up the cloud.

In the case of sunlight power generation, the atmospheric extinction, which is the sum of the effects of absorption and scattering in the atmosphere, is considered as an extinction proportional to the air mass, which corresponds to the thickness of the atmosphere. When OWPT is used at a transmission distance of about 10 km, which corresponds to that of the atmosphere, the same extinction is taken into account. Note that the atmospheric concentration varies in the direction of the thickness, but in the horizontal transmission direction parallel to the ground, the atmospheric concentration is almost constant, so the loss will increase even at the same transmission distance. When the propagation distance is of the order of 1 km or less, the transmission loss is sufficiently small.

Note that when the OWPT is used outdoors, weather conditions such as rain, snow, fog, and dust create large scattered light from visible size particles. Therefore, it is necessary to consider the visibility. Visibility is the maximum distance at which an object can be clearly seen by the human eye. Since OWPT also uses light, it should in principle be used under conditions similar to those that are visible to the eye. However, while the visibility of the eye only needs to reach a light power level that allows the object to be recognized, a sufficiently large ratio of the light power reached is desired for power transmission, so the applicable distance is expected to be shorter than the range of the visibility.

The above phenomena are mainly due to scattering by particles larger than the wavelength of the light. On the other hand, even if particle scattering does not result in transmission loss, regions with slightly different refractive indices will occur due to differences in density and temperature in the air. These cause fluctuations in the shape and phase components of the light beam, thus degrading the characteristics of the light beam. The air fluctuations consist of beam wander, which is a diurnal or seasonal effect of slow and relatively large air structure fluctuations, and scintillation, which is a fast fluctuation of small air structures. This has a significant impact on the beam-receiving conditions that fluctuate and when light is also used for wireless communication as well as power transmission.

5.3 COMPONENTS FOR OPTICAL WIRELESS POWER TRANSMISSION SYSTEMS

5.3.1 Configuration of Optical Wireless Power Transmission Systems

This section describes the typical configuration of the OWPT system, and the required functional elements and devices other than the light source and solar cells.

Although the OWPT is based on a beam light source and a solar cell, the actual system requires various other functional devices and subsystems, as shown in Figure 5.19. The system needs to detect and track the target device to be powered, a beam-scanning mechanism, control the shape of the light beam for efficient light irradiation, and communicate with the device via the power information. In addition, safety technology to prevent inappropriate irradiation of the light beam and a control system for multiple light sources and multiple terminals are also required.

The actual system configuration and required functions vary greatly depending on the application. As for the relationship between the light source side and the power receiving side of the OWPT, various combinations are possible, such as fixed light source to fixed device, fixed light source to mobile device, mobile light source to fixed device, and mobile light source to mobile device. In addition, the amount of power supplied, the transmission distance, the speed of the target device, and the application

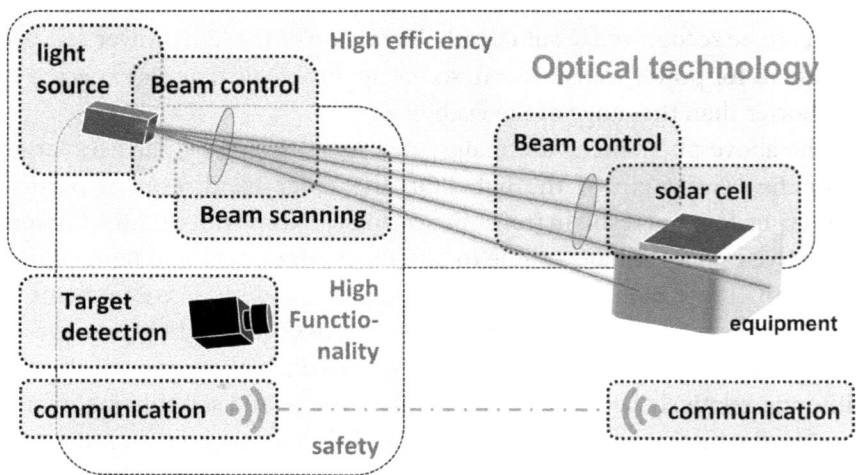

FIGURE 5.19   OWPT system configuration.

conditions such as on the ground, in the air, in space, underwater, etc., vary widely. The required technologies and features are different.

On the other hand, considering the configuration and functions as shown in Figure 5.19, most of the various functions can be applied from existing devices and technologies. In other words, OWPT is a very feasible technology. Some of the main important functions are described below.

5.3.2 Beam Irradiation Subsystem

If the beam spreads due to scattering, or if the beam spreads beyond the solar cell due to diffraction even without scattering, the beam collection rate on the solar cell will decrease and the power transmission efficiency will decrease. Therefore, if absorption and scattering are negligible, can high power transmission efficiency be achieved if all the light is incident on the solar cell with a sufficiently narrow beam width to avoid the effects of beam broadening?

In actual solar cell modules, such as silicon solar cells, the output voltage (open circuit voltage) of one cell is 0.5–0.8 V. Such low voltages are insufficient for actual load circuit operation or for efficient operation of the DC-DC converter circuit for boost voltage. In addition, the solar cell module is desired to output a high voltage to avoid losses due to wiring resistance in the circuit. Since the output voltage of a single cell cannot be controlled due to determination by material properties, multiple solar cells are usually arranged on a plane and connected in series. For example, a commercially available solar cell module of about 10 cm² contains 12 small cells and has an open circuit voltage of 6.9 V (0.575 V per cell) when connected in series.

Such a serially connected cell configuration works properly when light of nearly uniform intensity is irradiating on the solar cell module, such as in sunlight power generation or indoor lighting power generation. However, there is a problem with OWPT using light beams. Among the multiple cells, the cells that are not exposed to light irradiation are effectively insulators. Therefore, no current can be drawn from the solar cell module, resulting in zero conversion efficiency. Even if one of the cells is irradiated by weak light, the output current of that cell is small. Therefore, even if other cells can generate a large current, they cannot output power efficiently because they are limited to the cell with the smallest current due to the requirement of Kirchhoff's current law, i.e. the current continuity law in the circuit. In other words, the conversion efficiency is reduced. Although similar problems sometimes occur in existing solar cell applications, the OWPT must always anticipate such situations due to the use of beams.

One method is to introduce circuit countermeasures, such as applying bypass diodes to individual cells, as used in sunlight applications, however this causes excessive losses depending on the conditions and complicates the configuration of the solar cell module. On the other hand, if the light intensity distribution and the irradiation position are always fixed, it is effective to adjust the cell arrangement according to the beam intensity distributions. However, the need for fixation limits the application. As another solution, if the output voltage of the solar cells is high, series connection is not necessary, and all the cells can be connected in parallel if necessary to achieve the required cell area. This can be achieved by using multi-junction solar cells or short wavelength solar cells. However, these methods are limited in their operating conditions and increase the development effort for materials and devices.

As described above, in the case of OWPT using light beams, it is necessary to properly prepare the conditions and measures for irradiating the solar cell module with light beams. Although this seems simple, there are many challenges in achieving ideal characteristics.

The ideal conditions are (i) no light leakage from the solar cell module, (ii) irradiation of the entire solar cell module with uniform intensity, (iii) tolerance to variations in the shape and size of the incident beam, (iv) tolerance to variations in the transmission distance from the light source, (v) tolerance to variations in the incident beam position on the solar cell module, and (vi) tolerance to the incident beam direction. In OWPTs, since the light source side and the light receiver side are separated and have a high degree of freedom in their positioning, the system must meet these characteristics without incurring excessive losses.

The following methods can be considered to achieve uniform light by using passive optical (lens) systems.

The first is to use frosted glass or a light diffuser. However, the scattering of light returning to the incident side reduces the light utilization efficiency. This returning scatter component increases as the uniformity of the light is increased. Another problem is that random scattering cannot limit the area over which the light spreads.

The second is the use of a light shaping diffuser (LSD). Light is diffused by the structure of a microscopic and randomly sized lens array. By properly designing the lens shape, light can be distributed only over the required angular range, resulting in efficient and uniform irradiation. It can be thought of as a device that properly designs the function of a frosted glass. The use of an anti-reflection (AR) coating suppresses the reflective component and thus also increases the efficiency. However, as the angle of incidence of the incoming beam changes, the output direction of the beam also changes.

The third is the use of light pipes. This device uses a rod (pipe) structure made of glass to convert light with non-uniform intensity distribution into a specific shape, resulting in a uniform intensity distribution. It is usually applied on the light source side as a narrow rod structure. A lens is required to enlarge and project the output area of the light pipe to precisely irradiate the solar cells with uniform and required shape light. This scheme is analogous to using a light source of uniform intensity and appropriate shape. Of course, a light pipe could be installed on the solar cell side, but this would require a larger, longer, and heavier rod structure.

The fourth is the use of a fly-eye lens system [23]. A fly-eye lens is a lens array in which the lenses (element lenses) are arranged in two dimensions like the eye of a fly. The shapes of the element lenses are square and hexagonal to provide sufficient fill for efficient operation.

Since using only one fly-eye lens plate produces an image equivalent to the number of lenses, it is used to generate a large number of point light sources from a single light source. Its beam spread angle is determined by the focal length of the lens, which determines the spread range of multiple point sources. Array lenses are element lenses with the same characteristics, which can be said to be used in the same way as LSD.

If the beam shape and spread angle emitted from each element lens of the fly-eye lens are effectively the same, the light emitted from all the element lenses can be superimposed on a single spot. This can be achieved by combining a fly-eye lens and an imaging lens. Even though the light intensity incident on each element lens is different, relatively high uniformity can be achieved because all the outgoing beams are superimposed. However, with this configuration, the position of the image beam changes as the angle of incidence of the light changes.

Therefore, by combining two fly-eye lens plates and one imaging lens, there is no variation in the position of the image when the incident direction of the beam changes, i.e., it has the ability to always irradiate the solar cell position. The two fly-eye lens configurations and the experimentally generated image beams are shown in Figure 5.20.

In this figure, the element lens has a square shape, so the generated beam also has a square shape. When the incident beam is sufficiently larger than the element lenses, the incident beam is split into multiple element lenses that are superimposed to produce a highly uniform beam.

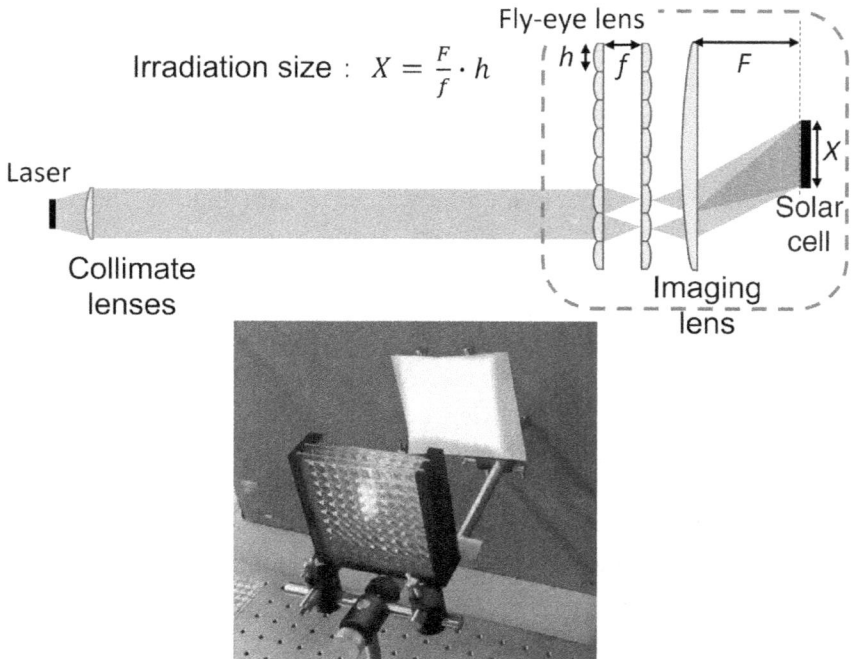

FIGURE 5.20   Fly-eye lens system configuration and demonstration of uniform rectangular beam generation.

This system is a well-designed lens system compared to frosted glass or LSD. Thus, the fly-eye lens system has (i) no light leakage from the solar cells, (ii) uniform intensity irradiation, (iii) tolerance to variations in incident beam shape and size, (iv) tolerance to variations in distance from the light source, (v) tolerance to variations in beam incident position [24], and (vi) tolerance to variations in beam incident direction. Since this system is a passive lens system, ideally there is no loss as a matter of principle by applying an AR coating to the target wavelength of the OWPT. Given the element lens height h, the element lens focal length f, and the imaging lens focal length F, the irradiation beam height H is calculated by the following equation.

$$H = \frac{h}{f} \cdot F \tag{5.16}$$

The disadvantage of this system is the thick (long) module size, which mainly depends on the focal length of the imaging lens. It is not possible to change to an extremely short focal length lens to achieve the required beam height equivalent to that of a solar cell module. This thick module size will limit the application.

In addition, there is a limit to the variation of the beam incidence direction. Its allowable angle of incidence θ_c is given by

$$\tan \theta_c = \frac{h}{2f} \qquad (5.17)$$

If the angle of incidence is greater than the allowable angle of incidence, the generated image position will not be on the solar cell. This allowable angle of incidence is, for example, 14.04°, as shown in Figure 5.20 for an element lens with a focal length of 20 mm and a size of 10 mm². This angle of incidence limitation is not a problem for applications where the light beams enter almost perpendicular to the fly-eye lens system. However, there is also a need for a method to increase the degree of freedom in OWPT applications.

Therefore, a configuration using a mirror as shown in Figure 5.21 has been proposed [25]. In the usual configuration, the fly-eye lens system irradiates light of the same shape to adjacent positions of the original

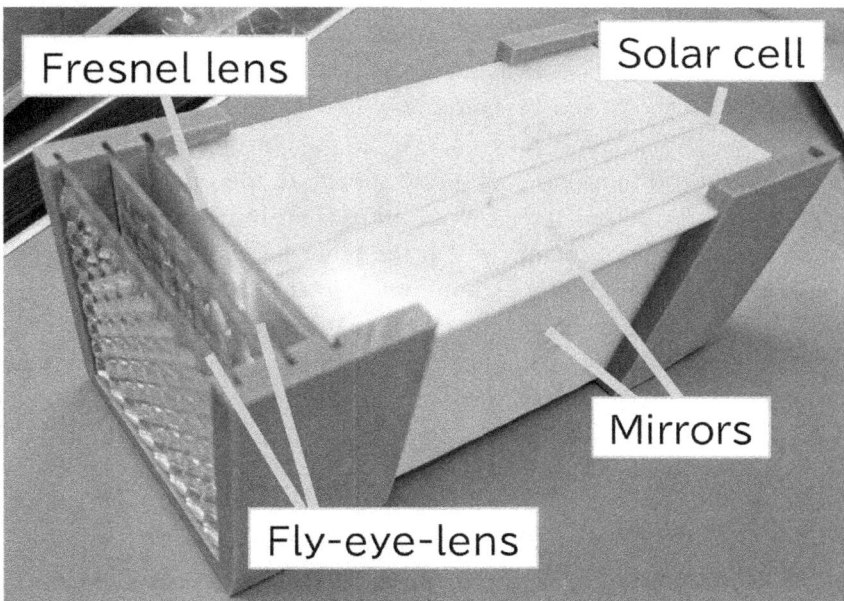

FIGURE 5.21   Fly-eye lens system with mirrors.

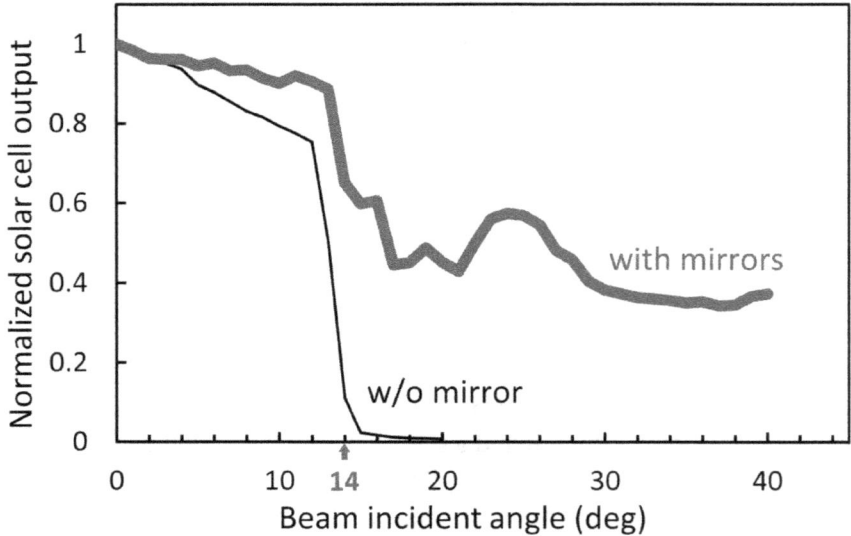

FIGURE 5.22   Measured incident angle dependent characteristics of fly-eye lens system.

irradiation position as the incident angle increases. By using the mirror, the irradiation beam is folded back. Even with a single fly-eye lens configuration or LSD, light leakage by reflection can be suppressed even when the incident angle changes by using a mirror. However, in this case, the light uniformity is reduced because the leaked light overlaps with the non-leaked light.

Figure 5.22 shows the measured angular dependence of the fly-eye lens system module with mirrors [25]. As a control experiment, a fly-eye lens system without mirrors is also shown. In the case of the module without mirrors, the solar cell output drops sharply around 14°, which is the allowable angle of incidence. On the other hand, the module with a mirror showed a sufficiently high output even at high angles of incidence above 14°. Note that although the output was expected to be close to that of the perpendicular incident condition, the output drops to about half of the expected value at high incident angles. The reason for this is thought to be that the lens characteristics were not designed to be optimal, and improvements are expected in the future.

5.3.3 Function Modules for OWPT System Configuration

In OWPT systems, beam-scanning mechanisms are important to ensure that the beam is accurately directed to the target equipment and solar

TABLE 5.2   Features of Various Beam-scanning Methods

	Scanning angle	Scanning accuracy	Scanning speed	Beam size/beam power
Direct drive (light source mounted on scanning mechanism)	++	o	–	++
Polygon mirror	++, 1-axis	o	–	++
Galvano mirror	o	o	o	o
MEMS	o	o	o	o
Liquid crystal devices	++	+	o	++
Electro-optic (EO) effect	o, 1-axis	–	++	++
Acousto-optic (AO) effect	–, 1-axis	–	++	+
Slow light beam deflection	++, 1-axis	o	++	o
Optical phased array (OPA)	o	–	++	–
Direction-dependent addressable multiple light sources	o	–	++	++

*   *Marks of comparison:* better ++ | + | o | – | poor

cells. Various beam-scanning methods have been developed and used for applications other than OWPT. For OWPT applications, these existing technologies can be applied, however, specific characteristics for OWPT are also required. The important characteristics for OWPT will be wide scanning angle range, high scanning accuracy/resolution, and large beam power/beam size. The advantages and disadvantages of OWPT applications are summarized in Table 5.2.

Although there are different methods, features, and characteristics, even for OWPT applications, the performance requirements vary from application to application. However, for many OWPT applications, the slower scanning speed of the beam may be acceptable compared to high-speed image frame generation, such as LiDAR, because it only needs to follow the moving speed of the target. For this reason, servo motors and galvanometers are considered important candidates.

As another function of the system, a target detection and recognition method is required to properly irradiate the light beam to the solar cells. For dynamic power transmission applications, high-speed and high-precision detection technology is required. There are two methods for recognizing targets and solar cell modules: passive markers for shape or image recognition and active markers using LEDs or other light-emitting devices.

For passive recognition, in addition to image recognition of the solar cells [26], methods for recognition of installed specific patterns such as

QR codes, and recognition of specific colors, are being considered [27]. As active markers, the use of continuously emitting LEDs [27], flashing LEDs [27, 28], near-infrared LEDs that are easily distinguishable from visible light [27, 28], and ultraviolet LEDs [27] that are easily distinguishable from outdoor sunlight have been reported. Another method that has been proposed to be robust to environmental differences is to use images of two different wavelengths of illumination. The light wavelengths are long-wavelength light near the absorption edge of the solar cells and short-wavelength light [29].

It is important to have an initial detection method to determine if there is any power-receiving device present in the room at all and where it is located over a wide area of the room. On the other hand, since the detection is aimed at irradiating the light beam, very accurate detection is required. In addition to position detection, orientation and shape detection are also required for proper beam irradiation when a fly-eye lens module is not used. Similar to detection, tracking after detection is also important. Different methods will be proposed in the future, depending on the application.

Communication technology is also required in an OWPT system. The power-transmitting side and the power-receiving side must at least communicate about the power transmission, such as the target detection status, the safety assurance status, the required power, etc., at the stage before the power transmission starts. Even after the start of power transmission, it is necessary to continuously communicate about the target detection status, the safety assurance status, the amount of power received, and the notification of the power transmission stop. In addition, communication as an electric power transmission management system, such as coordination between multiple power transmission system sides and multiple power-receiving sides, is also necessary.

In communication, it is necessary to communicate not only from the light source side to the power-receiving side, but also vice versa. In particular, it is important to transmit information from the power-receiving side to the light source side, such as power transmission requests, the status of the amount of power received, and the end of power supply. This communication must be wireless.

Since this is wireless communication, for the method using radio waves, existing methods such as Wi-Fi can be easily applied. On the other hand, if the use of radio waves is limited, communication using light will be a candidate. In this case, information can be superimposed on the power transmission beam from the transmitting side to the receiving side. Even when using a light source that is independent of the power transmission beam, developed light-based methods, such as Light Fidelity (Li-Fi) [30], can be applied. On the other hand, there is a problem with light communication from the receiver side to the power transmitter side. In the case of LED lighting, the energy required for communication may be large and the speed of communication may be limited in a method that irradiates light over a large area of room space, such as LED lighting. It is desirable to consider a method suitable for the application.

5.3.4 Safety Technology

The area in which the light beam for OWPT is transmitted may be entered by persons or equipment not related to the power supply, and there is a possibility of improper irradiation of the light beam. Due to the high energy density of the beam, safety in OWPT is an important issue that should be given the highest priority in the practical use of many applications, and requires a multi-faceted and proactive approach and countermeasures.

One possible application is to restrict the use of the OWPT system to areas that are inaccessible to people. Exposing equipment to high-power beams, even when no one is present, can cause overheating of the equipment or other effects. In this case, unnecessary light exposure to other equipment can be sufficiently suppressed, especially if the positions of the power transmitter and receiver are fixed with relatively short transmission distances. When power is supplied while the target is moving, as in long-distance transmission, the possibility of inappropriate beam irradiation increases, so accurate position detection and beam scanning are important [31].

When using light beams, it is necessary to consider not only the direct light beams used for the power supply, but also secondary light such as light leaking from the irradiation target, reflected light, scattered light, and so on. One W of laser light can be a problem in some situations. Although the proportion of this secondary light is small, 0.1% of a 1 kW transmission beam is 1 W, and it should be given appropriate attention. In practice, it is important to consider the condition in terms of light intensity density.

The beam safety criteria are explained here. For details and accuracy, it is necessary to refer to the appropriate information. The International Commission on Non-Ionizing Radiation Protection (ICNIRP) provides guidelines for maximum skin and eye exposure to radiation from lasers, LEDs, and lamps. The maximum permissible exposure (MPE) values are based on wavelength, exposure time (pulse width), intensity density, and other factors. The Japanese national standard JIS C 6802, which is based on the international standard IEC 60825-1, is used as the safety standard for laser products in Japan. This safety standard for laser products has seven classes of Class 1, 1M, 2, 2M, 3R, 3B, 4. The actual safety standards and necessary measures must be taken accordingly, as the regulations and laws of each country are also defined according to the products and the conditions of use.

Several safety techniques for OWPT have been reported. These include light curtains with sensing beams around the power transmission beam [32], controlling laser emissions within Class 1 for very short exposure times based on specific beam generation methods [33], using eye-safe wavelength band beams [34], using LEDs that are slightly safer than laser beams [35], and using camera images to detect unsafe conditions [36].

The following is an example of a camera-based safety system that has potential for a variety of applications. Figure 5.23 shows the basic concept of a camera-based safety system. The camera is installed on the light source side and can detect approaching objects through image recognition. When a sensor detects an object approaching or entering the safe distance, it feeds this information back to the light source, which stops the light before the object touches the light beam. The safe distance between the light beam and the object needs to be considered for a definition, as human psychology and physical safety differ, but in an indoor environment, a safe distance of 20–30 cm is required when the object speed is limited to 5 m/s or less [37].

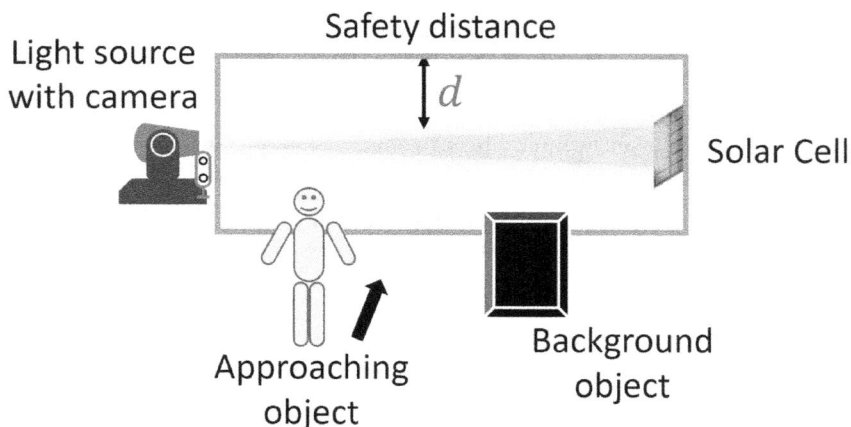

FIGURE 5.23   Schematic configuration of a camera-based safety system.

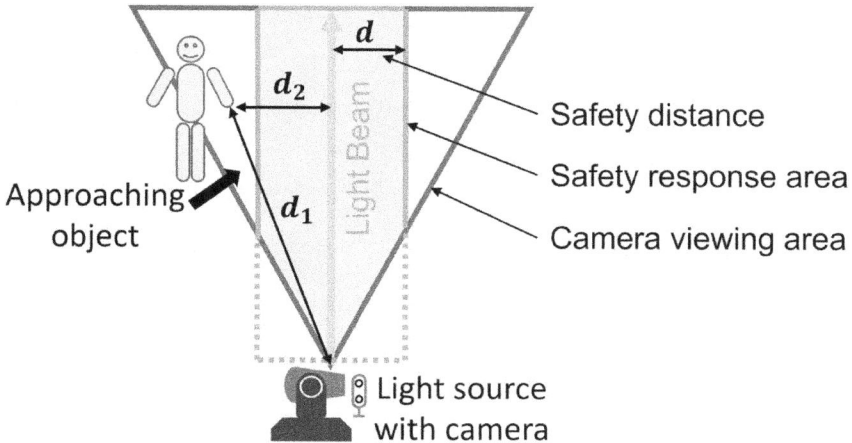

FIGURE 5.24   Camera viewing angle and ideal safety area.

Figure 5.24 illustrates realistic conditions for the safety system. For an ordinary two-dimensional camera with a fixed viewing angle, an image of a triangular area is obtained. Even if an object approaches this triangular area, the distance from the object to the light beam depends on the distance between the light source (camera) and the approaching object. This may not be safe or may turn off the light unnecessarily. Therefore, it is necessary to define a rectangular safety area. In such a rectangular safety area, distance information between the light source and the approaching object is essential. Possible methods for obtaining 3D (depth) information include the dual camera method, the structured light method, and the LiDAR method. These 3D cameras have become popular in recent years and are easy to use.

Note that in order to achieve a high level of safety, it is necessary to use not only one safety system, but several methods.

5.4 APPLICATIONS OF OPTICAL WIRELESS POWER TRANSMISSION

This section introduces the possible and expected application areas for the OWPT system with some reported examples.

The application areas are shown in Figure 5.25. OWPT can be applied to power ranges from mW to kW and distance ranges from cm to km. In other words, it is applicable to power most existing equipment that use electricity. In addition, because it offers the new capability of being wireless, it can be applied to systems that have never used electricity before. This is similar to

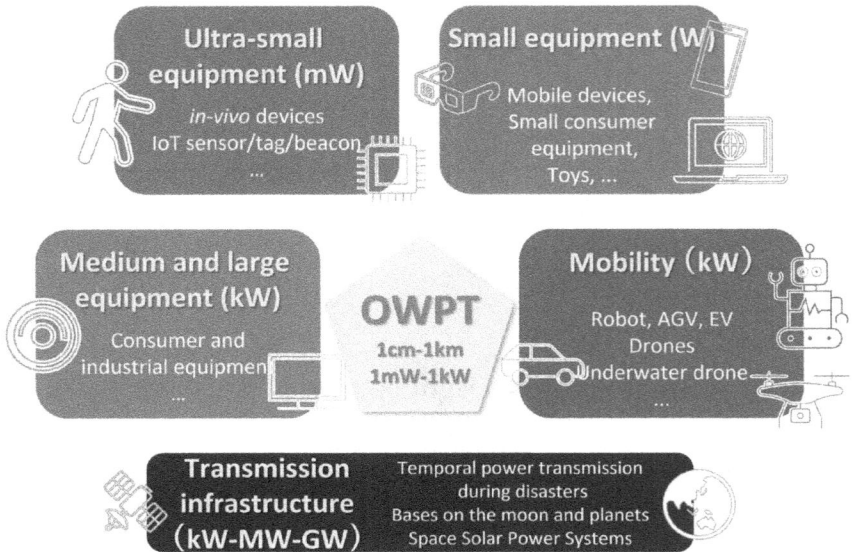

FIGURE 5.25   Applications of OWPT.

the concept of the Internet of Things (IoT), which provides sensing as the main function of various equipment, however, OWPT can provide not only sensing but also functions such as high-power actuation.

The classification shown in the figure covers very small devices in the mW class, such as IoT sensor terminals; portable terminals, such as toys and smartphones, ranging from a few watts to tens of watts; consumer devices; and medium and large equipment in the kW class, such as electric vehicles (EVs), drones, and robots. It is also effective for applications inside equipment where wiring is problematic, such as rotating and moving mechanisms inside equipment. These applications are also being considered for existing wireless power transmission methods, but the long-range transmission of OWPT is advantageous in that it can significantly reduce the number of light source devices that serve as infrastructure equipment. In addition, the feature of not generating EMI would expand the applications and their conditions. In terms of achieving practical efficiency, OWPT is the only method for dynamic charging of mobility that moves three-dimensionally in air, water, and space.

The use of the system as a power transmission infrastructure remains a challenge due to its low power transmission efficiency. However, it could be used for temporary power transmission or when no other alternative method is available.

5.4.1 In vivo Terminals, Small IoT Terminals, and Consumer Applications

OWPT is considered to be effective for body-implanted terminals. i.e. in vivo terminals, due to the difficulty of wiring to the outside of the body. However, since light from outside the body does not reach deep inside the body, it is expected to use low electrical power and to supply power to a solar cell placed just under the skin. Applications have been reported for use in cardiac pacemakers that do not require surgery to replace batteries [38,39] and for remote recharging of microchips about 1 mm^2 implanted in the body [40–42].

The concept of small IoT terminals of a few centimeters square, such as sensors, tags, and beacons, is also promising as an OWPT application. Small IoT terminals are likely to be used in large numbers, but the heavy workload of wiring installation, battery charging, and battery replacement has slowed the spread of these applications. OWPT is still a charging method, but since multiple terminals can be charged remotely, the workload can be greatly reduced.

For this reason, two configurations have been reported: one for remote charging of terminals from a worker, robot, or drone, as in the flashlight in Figure 5.26 [43], and the other for OWPT from a fixed scanning light source to multiple terminals sequentially, as in Figure 5.27 [44]. The former is proposed as an LED light source system, which is advantageous in terms of safety [45,46]. This LED light source system is reported to provide an output of more than 1 W at a distance of 1 m from a 5 cm^2 solar cell [47].

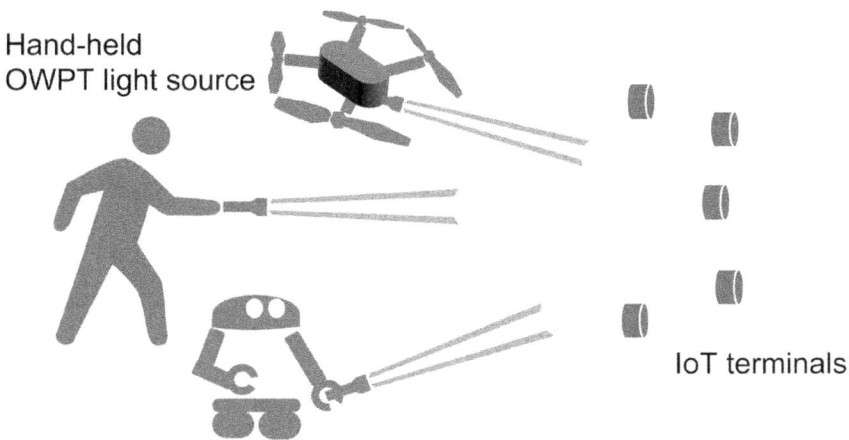

FIGURE 5.26   Flashlight-type OWPT charging to small IoT terminals.

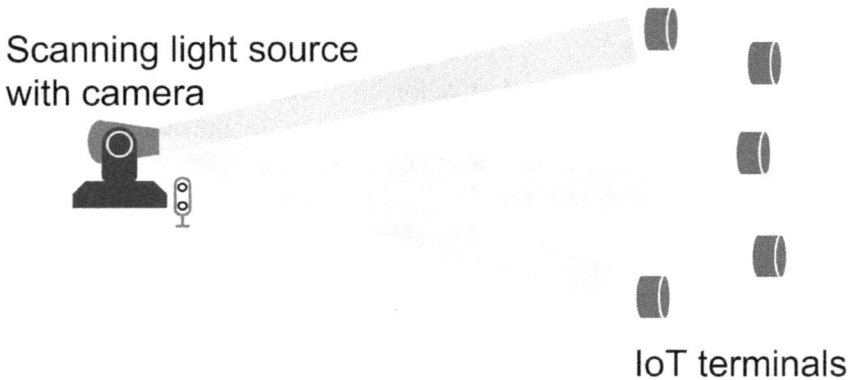

FIGURE 5.27 OWPT charging to small IoT terminals using a scanning-type light source.

Due to the monochromatic light effect, the power corresponds to a light density twice that of sunlight irradiation. It should be noted that the wide beam spread and low light intensity density of an ordinary flashlight make it difficult to use for charging the battery.

On the other hand, the latter is proposed as a semi-automatic sequential OWPT system for multiple small IoT terminals installed indoors [44]. In such applications, when the exact location and number of installed terminals are not known, it is difficult to obtain such information automatically. Therefore, the initial locations of the terminals are set manually by visually observing the state of light irradiation and the images from the movable camera. When power is supplied, the system automatically irradiates light beams sequentially in the set direction for the set time. Considering the low accuracy of the prior manual settings, the system performs high-precision position detection and beam irradiation by image recognition at the time of power supply. In the report, the actual configuration consists of a light source and a 3D camera mounted on a two-axis servo motor type directional beam scanner, which can irradiate light beams at a range of 540 degrees in the plane and 190 degrees vertically.

Note that, in this method, multiple terminals are charged sequentially using a small number of light sources. Since the reported example requires about 1 second for beam direction scanning and high-precision position detection, the light source utilization efficiency decreases as the number of terminals increases. In addition, the terminals also require a small amount of battery because they are not continuously supplied with power, but if the light irradiation time for each terminal is long, the unsupplied time

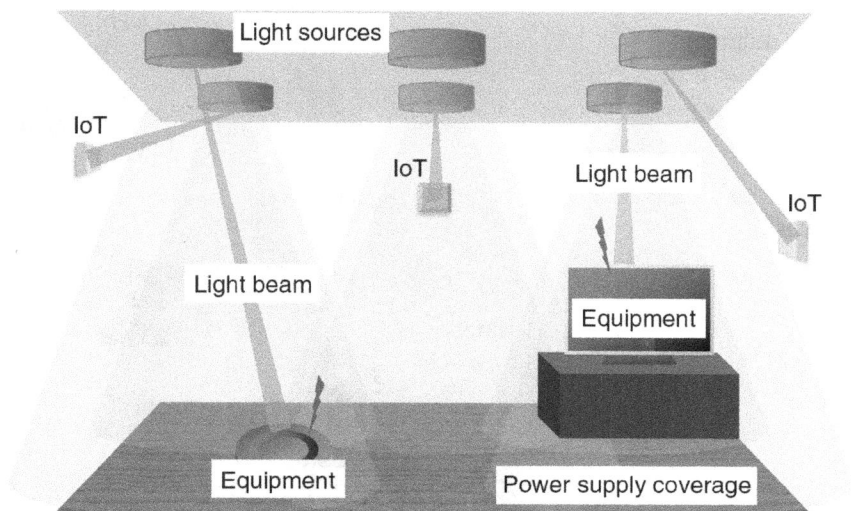

FIGURE 5.28   Application scenario of indoor OWPT.

for each terminal increases, resulting in a large amount of battery use. It is important to shorten the light source control time and the system operation algorithm that takes the battery load into account.

Figure 5.28 shows a picture of an indoor OWPT. Indoor applications such as information terminals and home appliances can be said to have the same configuration as IoT terminals, but in addition to increasing the amount of power supplied, it is also necessary to address the movement of the terminal position and safety. Since the degree of freedom of the terminal position is increased, advanced mechanisms are required to reduce the occasion of light beam shielding and other measures. For this reason, as shown in Figure 5.29, a method has been proposed in which multiple light sources are installed and the more effective light source is sequentially selected based on detection and distance measurement by a 3D camera [26]. Since increasing the number of light sources limits their applications, the use of directional mirrors instead of directional light sources has also been proposed [48]. If necessary, the beam output from the light source is reflected by mirrors to irradiate the target.

In consumer applications, the solar cells on the surface of the terminal will constrain the design of the terminal, such as the color. Since OWPT uses monochromatic light, the wavelengths used for power supply are transmitted to the solar cell and the other wavelengths can be used to control the color of the design [49]. Such technologies will be important for

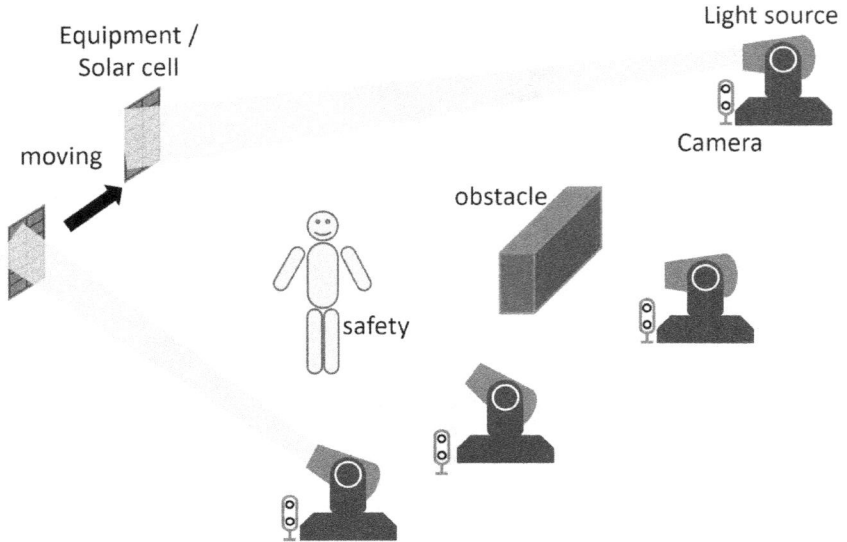

FIGURE 5.29 Multiple light sources for continuous power transmission and safety.

consumer applications of OWPT. On the other hand, the application of visible light lasers to lighting systems is also being considered, and an OWPT using visible light lasers for lighting has been proposed [50].

Due to the wide range of applications for such consumer and IoT applications, several companies are actively working on it. Wi-Charge, a venture company in Israel, has pioneered such OWPT systems for consumer applications [33]. It is targeting information terminals and small power devices up to a few watts. However, the technical details of this company's system have not been disclosed. According to the company's patent application, retroreflective mirrors are used on both the light source side and the power-receiving side as the two mirrors that make up the laser cavity. A solar cell is installed under the mirror on the power-receiving side. By enabling laser operation with retroreflective mirrors, the patent proposes an interesting mechanism that automatically detects the position of the target (the mirror on the power-receiving side) without the target recognition system, automatically scans the beam to the target without explicit scanning mechanism, and immediately turns off the laser beam if an unwanted object enters the beam. Similar technology has also been developed in reference [51]. Another venture company in the United States, PHION Technologies, and a telecommunication company in Japan,

Softbank [52], are also conducting research and development on OWPT systems for consumer, industrial, and IoT applications.

5.4.2 Dynamic Charging for Mobilities

Mobility has been developed with fossil fuels and engines, but low thermal energy efficiency and configuration complexity have been problematic. The inefficient use of fossil fuels in mobilities has been a major contributor to CO_2 emissions, and the complexity of engine mechanisms has not only affected cost, but also made it difficult to install multiple engines, especially if the mobility is low, limiting the characteristics, functions, and applications of mobility. Electrification has been promoted to solve these mobility problems. Electrification is an effective means of reducing CO_2 emissions because there is significant room for improvement in system efficiency. In addition, since motors are simpler to construct than engines, it is possible to reduce cost and size, and increase performance and functionality by installing multiple units. This has led to the development of different types of mobility.

This electric mobility requires the installation of a battery. While these batteries support the development of mobility, they need to be recharged or replaced, and their low energy density (about 0.2 kWh/kg for lithium-ion batteries) limits the distance that can be traveled. Although engines also need to be refueled, the work is completed in a short time, and the high energy density of gasoline (about 12 kWh/kg) allows long-distance travel.

Due to the low energy density of batteries, EVs and drones have a large battery to total weight and cost ratio of 20–50%, and adding more batteries will not increase the travel distance because of the energy used to move the batteries. In addition, continuous automatic operation without human intervention is desired in the near future, but charging work and charging time will hinder this, and even if battery characteristics are further improved, the increase in charging time and charging power will become a problem.

Therefore, the application of OWPT to mobilities is attractive, as shown in Figure 5.30. The ability to supply power wirelessly from a distance simplifies the charging process, and especially in the application of dynamic charging, i.e., in-motion power transmission, there is no downtime due to recharging [53]. In this case, ideally there is no need to install the battery, and realistically a small battery size will be sufficient. In addition, the battery load can be reduced by eliminating the need for intensive fast charging.

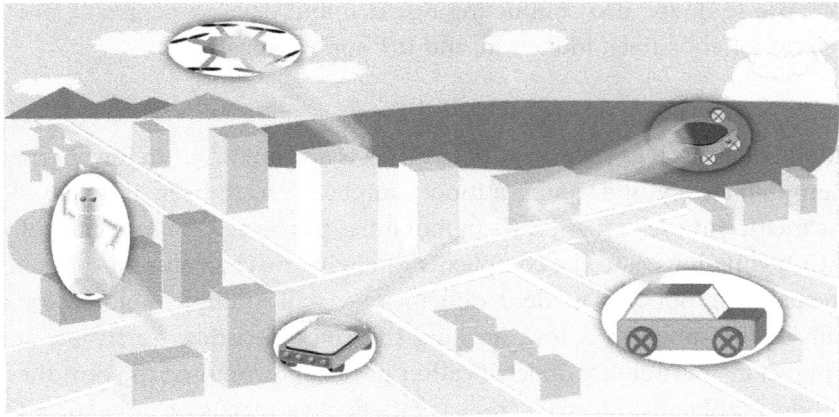

FIGURE 5.30   OWPT to various types of mobilities.

Thus, OWPT has many advantages for use in mobility applications, however, it also has drawbacks. In this section, we consider the main issues of OWPT from the perspective of mobility applications.

First, there are concerns about the increase in CO_2 emissions for electricity power generation due to the low efficiency of power transmission. However, it is pointed out that OWPT dynamic charging of EVs is beneficial because it enables a reduction in the amount of batteries installed in EVs, thus reducing the large amount of CO_2 emissions during battery manufacturing [54]. As a detailed evaluation of the CO_2 emissions in the lifecycle of EVs, it has been reported that even a power transmission efficiency of 16% is effective in reducing CO_2 emissions for some limited use cases of EVs, as shown in Figure 5.31, and that further improvement of power transmission efficiency can contribute to CO_2 reduction in many use cases [55]. Although the actual realization of optical wireless power transfer while EVs are in motion requires the development of many light source system infrastructures, and it is not an easy shift from the current EV policy of solving problems by improving battery performance, it is worth considering from the perspective of CO_2 emission reduction effects.

OWPT can provide significantly different values from those of the past, such as the possibility of reducing the work of power transmission and batteries. It is effective not only for EVs, but also for ground-based mobile robots, cargo-carrying robots called auto-guided vehicles (AGVs), flying drones, and underwater robots/drones. Some examples of OWPT

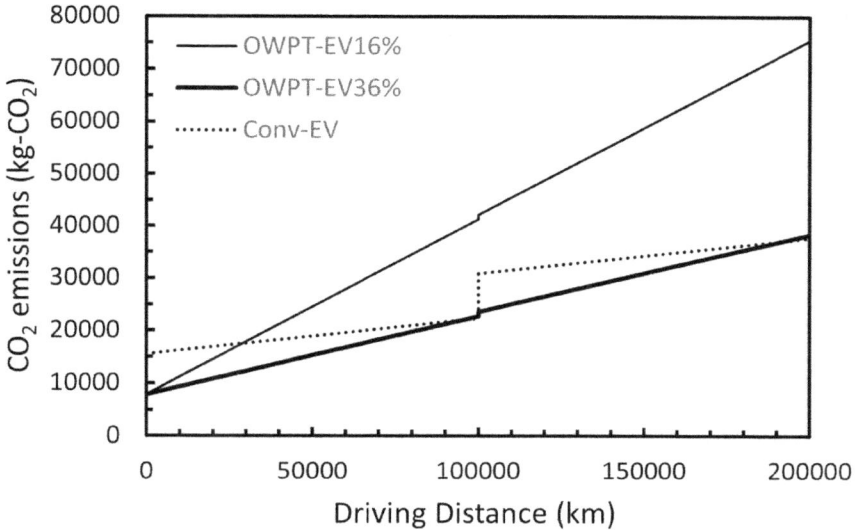

FIGURE 5.31 CO_2 emissions over the life cycle (total driving distance) of EVs. Standard type OWPT-EV equipped with 10% of the battery capacity of a conventional EV.

Source: Modified from [55].

experiments that assume actual mobility and mobility-aware approaches are summarized in Table 5.3, although there may be other related efforts.

This table shows that cases began to be reported in the 2000s; the NASA case in 2003 is considered the first report of powering mobility. In this case, solar cells were suspended from lightweight model airplanes, and the direction of the light beam was manually controlled for light irradiation, as seen in the following video (www.youtube.com/watch?v=InuP RZjwykU).

In the mid-2000s, the group of Prof. Kawashima of Kinki University and Hamamatsu Photonics flew a toy kite plane and a small helicopter (multicopter drone) [56,57]. These were flown using high-power fiber-coupled lasers in the 500 W class. They demonstrated continuous flight of a drone weighing more than 1 kg for several 10 minutes. By attaching a retro-reflective prism to a solar cell module hanging from the drone, the reflection of the light beam for power supply was used for position detection. The drone was also equipped with a camera and transmitted images wirelessly via radio waves as shown in the video (www.youtube.com/watch?v=qMLB Amqj288).

TABLE 5.3   Reports on OWPT for Mobility

Year	Mobility type	Affiliation	Results
2003	Toy model plane	NASA Armstrong Flight Research Center	Infrared laser OWPT to a model plane
2006	Toy kite plane	Kinki University and Hamamatsu Photonics	Long time flight of a kite plane with 350 W laser beam in a dome stadium
2008	Multicopter drone	Kinki University and Hamamatsu Photonics	Using a 580 W fiber-coupled LD, a multicopter drone with about 1 kg and a diameter of 1 m is flown for a long time in a dome stadium
2010	Multicopter drone	LaserMotive (now PowerLight Technologies)	12.5 hours of continuous flight of a multicopter drone
2016	Toy micro-car	Nagoya University	Laser irradiation to a toy car from above the roof at a fixed location on the course
2017	Toy micro-car	Miyazaki University	Power supply to a toy car from LEDs installed on the course surface
2018	Toy micro-car	Kanazawa University	Continuous driving of a toy car by beam scanning from outside the course based on position detection
2018	Miniature flying robot	Washington University	Infrared laser power beaming to a flying micro-insect robot
2018	Toy train	Wi-Charge	Power supply to multiple toy-trains from a ceiling-mounted light source
2019	Toy drone	Tokyo Inst. Tech.	Dynamic charging to a toy drone using VCSEL for a few cm floating
2020	Toy car	Tokyo Inst. Tech.	Dynamic charging to a toy car using VCSEL on the straight part of an oval racetrack

In addition, regarding flight by optical wireless power transmission for drones, around 2010, a group led by a venture company, PowerLight Technologies, in the United States demonstrated continuous flight of a multi-copter type drone, also several tens of centimeters in size, for as long as about 12 hours [58]. The demonstration can be seen on video (www.yout ube.com/watch?v=8hhv9Cu98us).

On the other hand, ultra-compact drones weighing less than a few hundred grams have become increasingly powerful in recent years. If such ultra-small drones can achieve continuous flight, applications such as sensing and imaging will expand. Smaller drones require less power, but also present difficulties because their limited weight limits their ability to carry additional equipment. As shown in Figure 5.32, even an ultra-small

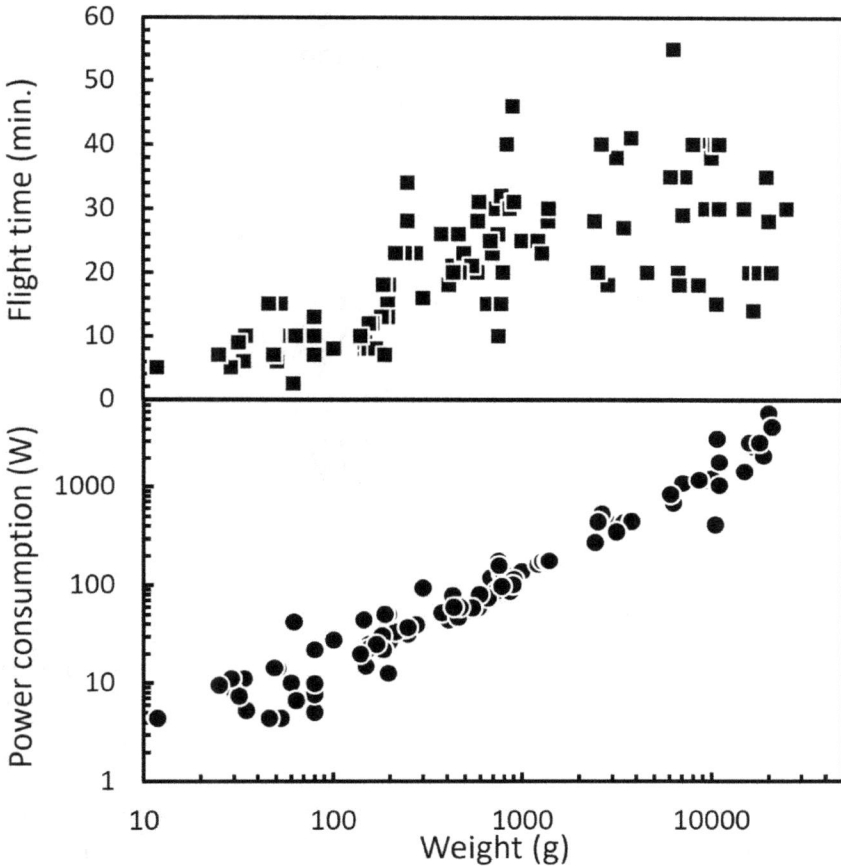

FIGURE 5.32 Power consumption and flight time vs. weight for many commercial drones.

drone weighing about 100 g, including toys, consumes several watts or more of power, requiring tens of watts of optical power when OWPT is applied. Note that this class of relatively high light power is required even to operate the toy. In fact, the continuous floating of an ultra-small toy drone with a light output of 20–30 W has been confirmed in Figure 5.33 [44,59]. In this case, the solar cells are installed on the ground and the wires are connected to the toy drone. The reason for limiting the system to vertical floating only is to reduce the development of the target detection and beam-scanning functions.

Next, examples of OWPTs for ground-based mobility are described. Ground-based mobility includes EVs, robots, and AGVs, as mentioned

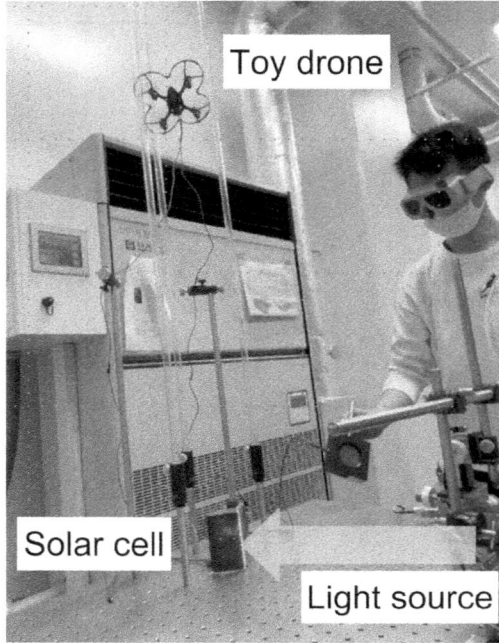

FIGURE 5.33   Floating of an ultra-compact toy drone by OWPT. As an initial test, solar cells are installed on the ground and power is supplied to the drone via wiring.

earlier. These require nonstop and continuous operation with automatic operation, but the work of recharging batteries, charging time, and traveling to recharging locations significantly reduces the utilization rate of the equipment. Although OWPT requires the provision of power supply facilities as infrastructure, the impact on installation costs can be minimized because the infrastructure is shared equipment for many mobilities.

Similar to the examples of dynamic charging of flying mobilities, there are not many detailed studies of dynamic charging of ground-based mobilities. As listed in the table, some conditions of light beam installation have been demonstrated by university groups using micro-toy cars. A case study is presented here, where OWPT is used from behind (in front of) the moving toy car, taking advantage of the long-range transmission of the light beam with its small angle of spread. By also scanning narrow angles of the beam, the number of light sources can be significantly reduced. In Figure 5.34, a solar cell is mounted on a toy car. Instead of a battery, an electric double-layer capacitor (EDLC) with a large capacitance of about

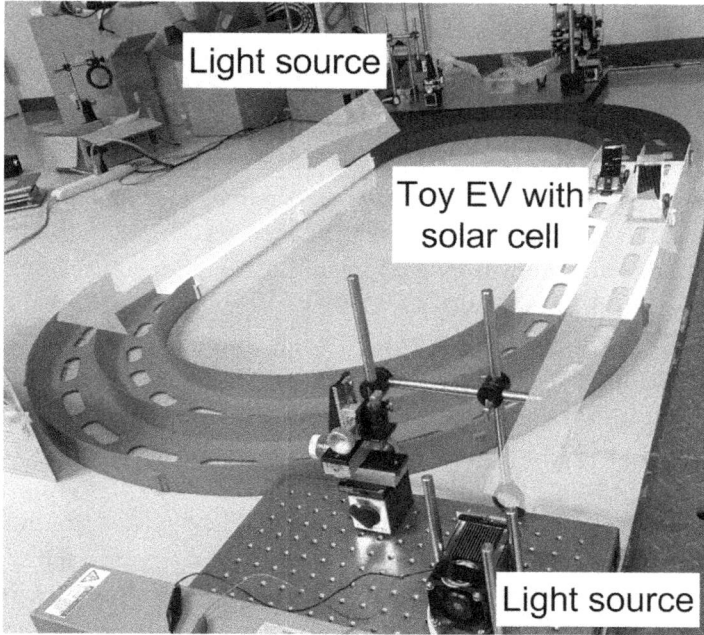

FIGURE 5.34   Running a toy car with OWPT dynamic charge. The toy car is set with solar cells and powered by two laser beams on a straight course.

1 F is used to operate the toy car in areas where there is no light irradiation. In this demonstration, two light sources, one with a light output of 20 W and the other with a maximum of 40 W, are installed on two straight sections of the oval track. With the charged capacitor alone, the toy car ran for approximately two laps, however, with the two light beams, two additional laps were observed [44,60]. By increasing the light output and further increasing the light beam utilization ratio, continuous running will be possible while including non-irradiated sections.

Although the actual application of OWPT dynamic charging to electric vehicles with high weight and large size is not easy, considering the scaling law (first-order approximation) of system characteristics according to weight and size, it is expected that existing laser light sources with several kW output power could be used for operation.

In addition to the above consideration of OWPT equipment configuration and CO_2 emission reduction, it is also important to develop the infrastructure for such a power source for EV dynamic charging. Considering the long-distance nature of OWPT, light sources need to be installed at intervals of 100 m to 1 km. Considering Japanese roads, for example, several hundred thousand to several million light sources would be required. In addition, since multiple light sources are needed to supply power to multiple vehicles at the same time, an enormous number of light sources would be required. However, considering that Japan has 80 million vehicles and 3 million street lights, this number is considered feasible. Since the light sources would be installed in a similar manner to street lights, there would be no extra burden on the road surface and road maintenance.

Fortunately, independent of the OWPT system, the installation of solar cells on EVs has begun to be put into practical use to achieve an auxiliary function [61] for charging by sunlight, and as an extension of this method, OWPT dynamic charging is also being considered.

5.4.3 Underwater Applications

Underwater and offshore applications are closely related to the social economy such as primary industries; energy such as hydroelectric power, offshore wind power, and floating solar power; infrastructure such as ports, bridges, communications, and transmission lines; disaster relief; and environmental and consumer applications. In these applications, it is important to build new systems and services that support a sustainable society. However, radio waves do not penetrate underwater, making it impossible to use the wireless technologies that have helped develop applications on the ground, in the air, and in space applications. In practice, sound waves have been used as a wireless technology, but their characteristics place many limitations on their use and performance. In terms of power transmission, until now, both fixed and mobile devices have been powered underwater by cables or batteries. The difficulty of using cabling and charging underwater has been greater than on land, resulting in system and service limitations.

Blue and green light transmitted underwater are promising as underwater wireless technologies, thanks to recent improvements in light source performance. Due to the difficulty of wireless power transmission by other methods, OWPT is promising as the only method for remote wireless power transmission underwater.

Table 5.4 shows examples of reports on underwater OWPT. Only about 10 institutions worldwide have conducted research on OWPT. The number of papers is very limited, at about 20, and the reports are from 2015 onwards. This does not mean that underwater OWPT is less important, but rather that the number of OWPT studies remains low.

TABLE 5.4   Reports on Underwater OWPT

Year	Organization	Results
2015	JAMSTEC	Wireless charging concept using LEDs [62]
2015	JAMSTEC	Proposal of AUV underwater charging station
2018	Kyungsung University	Experiments of 100 mW light output and 4 m of tap water and seawater [63]
2019	Tokyo University of Science	Initial study of underwater optical communication and power transmission
2019	Ozyegin University and Aristotle University	Study of underwater optical communication and power transmission (SLIPT)
2019	Kyungsung University	Design of underwater visible light power transmission and communication system
2019	Kanazawa University	Optimum wavelength analysis according to underwater transmission distance [64]
2019	Kanazawa University	Propagation experiment of 50 mW and 10 cm [65]
2020	Tokyo Inst. Tech.	Experiment through water waves between above water and under water [66,67]
2020	Ibaraki University	Theoretical analysis of underwater visible light power transmission and communication [68]
2020	Kyungsung University	Theoretical analysis of wavelength considering absorption coefficient of water [69]
2021	Chiba Inst. Tech.	Power feeding experiment in 90 cm seawater [70]
2021	Fudan University	3.96 mW and 2.3 m power transmission experiment [71]
2021	Southeast University	5 m power transmission experiment [72]
2022	Chiba Inst. Tech.	Power transmission experiment in 90 cm of seawater [73]
2022	Tokyo Inst. Tech.	Power transmission experiment of 0.76 W output at 90 cm [25]
2022	Tokyo Inst. Tech.	Power transmission experiment of 0.81 W output at 9.9 m [74]

Most of these reports investigated the basic OWPT configuration of the light beam passing through the water and received by a solar cell, and the light output was low, at about 100 mW or less. In practical applications, several watts or more will be required to operate underwater equipment. When considering OWPT in water, it is important to consider that propagation losses are greater than in air, that the effect of suspended particles is more likely to occur, that currents and high water pressure must be dealt with, and that water waves exist.

Light loss in water is caused by absorption by water molecules and scattering by particles. Figure 5.35 summarizes the spectrum of water loss

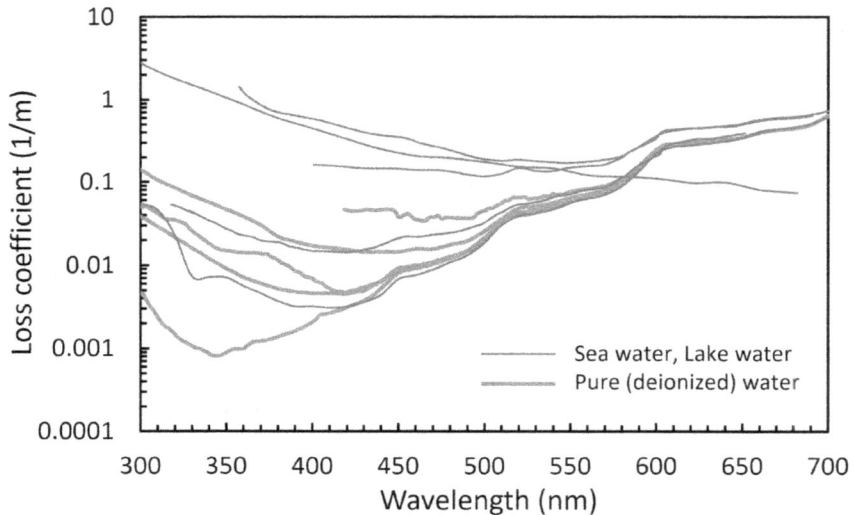

FIGURE 5.35 Loss in water. Various reported results are shown for pure water, and sea/lake water. The graph is modified from [74].

coefficients from various reports. It can be said that there is little loss in the wavelength range from blue (400 nm) to green (550 nm). Although the loss coefficient varies depending on the water in the natural environment and the quality of the water, the maximum distance for underwater optical wireless power transmission is considered to be 10–100 meters. This distance can be said to be slightly shorter than the visual distance.

Considering the configuration of the actual OWPT system, the maximum efficiency of blue semiconductor lasers has reached 50%, and several W to kW have been achieved as a light source system. On the other hand, solar cells on the light-receiving side have problems. Solar cells suitable for blue light have not yet been realized. For this reason, existing solar cells have been used in experiments, but since they are suitable for near infrared, their efficiency drops to less than 20% for blue light. As mentioned above, short-wavelength solar cells are not only efficient in principle, but also essential for underwater applications.

The other functional elements of the OWPT are not significantly different from those on the ground. As a research case study, an example of W-class underwater OWPT has been reported using a configuration based on a fly-eye lens system, as shown in Figure 5.36 [25].

The experiment was conducted in a 90 cm long acrylic water tank filled with tap water. The light source and the solar cell module were placed

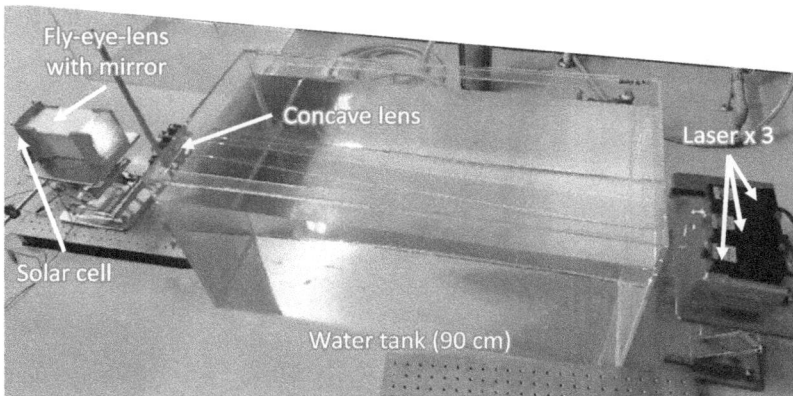

FIGURE 5.36   Demonstration of underwater OWPT using 90 cm long water tank.

outside the tank. Since the actual application will place devices inside the container, the experiment is a suitable simulation. A laser with a wavelength of 450 nm and a maximum light output of 6 W per laser source was used as the light source, and GaAs solar cells with a fly-eye lens system were used on the light-receiving side. The light-receiving area of the fly-eye lens module is 8 cm², and the solar cells are also 8 cm². A total light output of 17.5 W from three light sources and solar cell output of 2.25 W were obtained by using three 6 W light sources [74]. Since a fly-eye lens system is used, an error of a few centimeters is allowed in the incident position of the three beams.

In addition, highly reflective mirrors were placed inside a water tank to repeatedly transmit light back and forth, achieving a maximum transmission distance of about 10 meters [74]. Due to the loss of light in the water, the solar cell output was 0.81 W, although the light output was as high as 17.5 W. Increasing the light output is possible, and a 10 m-100 W class power transmission is expected in the near future.

In the case of underwater OWPT only, as described above, the light loss is significant even at a distance of about 10 meters. Even in highly transparent water, the maximum distance for power transmission would be about 100 m or less. In reality, it is not easy to make water transparent in oceans, lakes, and swimming pools. Therefore, a configuration that combines not only between water but also between the air and underwater would be important for OWPT in various situations. As shown in Figure 5.37, in addition to emitting a beam from underwater to above water, and re-injecting it into water, several other configurations are being considered,

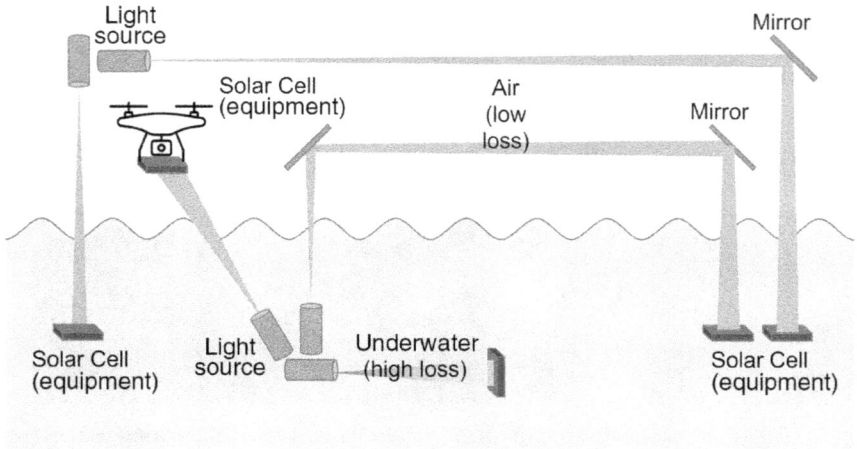

FIGURE 5.37   OWPT combining underwater and above water.

such as from above water to underwater and from underwater to above water [66].

In the actual propagation of a light beam between underwater and above water, water waves are considered to be a problem. Since the refractive index of water is about 1.33, which is close to that of air, the normal incidence reflection at the air interface is negligible, at about 2%. However, the water wave is tilted and refracts the light beam at the interface. Although this is effectively a scattering phenomenon, unlike particles, the size of the wave is large enough to change the light beam in a variety of directions. This results not only in a broadening of the beam, but also in a continuous time-varying intensity distribution on the solar cell surface.

Various future studies will provide an effective system for underwater OWPT.

5.4.4 Space Applications

Space applications of optical wireless power transmission are also being studied. No actual experiments or demonstrations have yet been carried out in space. The largest project is space photovoltaic power generation. Solar cells installed in space can generate electricity continuously throughout the year. Wireless power transmission is being considered to deliver the generated power to Earth. Most of the studies are based on microwave wireless power transfer, but the Japan Aerospace Exploration Agency (JAXA) is also considering laser power transfer [75]. The use of lasers allows the use of narrower beams, reducing the area of the light-receiving system on the ground. One problem is that transmission efficiency varies with weather conditions.

Further applications are expected in space. The extremely high cost of transporting materials into space limits the use of even readily available wiring on the ground. For this reason, the OWPT system can be used to replace wiring inside the space station and to provide power between multiple bases that will be built on the Moon and elsewhere in the future.

In addition, rovers on the lunar surface, which are desirable for long-term exploration in the shaded areas of craters, can be operated for long periods using OWPT to obtain power from solar facilities around the craters. Kinki University, which introduced the dynamic charging using OWPT to a drone, has also been working on such OWPT for rovers on the Moon [76].

5.4.5 Optical Power Transmission Using Fiber Optics

Optical fiber power transmission (power over fiber; PoF), a technology similar to OWPT, is described here. A light beam is injected into an optical fiber, and the light energy is converted into electrical power by a solar cell installed at the output end of the fiber. In other words, the transmission medium is changed from air or water in OWPT to optical fiber. Of course, PoF is wired, not wireless.

There are several advantages to using a fiber. Compared to copper cabling, the use of insulating materials in the optical fiber provides an insulated connection and provides surge resistance against lightning strikes. Optical fiber communications can be combined with PoF using a single cable. Fiber is lighter than copper cabling. Compared to OWPT, power can be supplied even when without line of sight. The light collection rate is high even over long distances of several kilometers, including with a small receiver, because the beam is not spread by diffraction. There is high safety because the light beam is confined within the optical fiber.

Because of these advantages, this method is already being commercialized for low-power applications. The high-efficiency solar cell described in the solar cell section was developed for this PoF. By using a special optical fiber, a power output of 40 W or more has been reported at the end of a fiber with an optical input of 150 W and an optical fiber length of 300 m. In this experiment, optical communication was also transmitted simultaneously

[77]. OWPT and PoF are expected to expand in the future as methods of power transmission using light.

REFERENCES

[1] T. Miyamoto, "Optical Wireless Power Transmission Using VCSELs," *Proceedings of SPIE*, vol. 10682, pp. 1068204, 2018.

[2] W. J. Robinson, Jr., "The Feasibility of Wireless Power Transmission for an Orbiting Astronomical Station," *NASA Technical Memorandum*, X-53701, 1968.

[3] K. Horii, "Consideration of Power Transmission Using Laser Beam," (in Japanese) *Bulletin of the Electrotechnical Laboratory*, vol. 36, pp. 279, 1972.

[4] W. Shockley and H. J. Queisser, "Detailed Balance Limit of Efficiency of p-n Junction Solar Cells," *Journal of Applied Physics*, vol. 32, pp. 510, 1961.

[5] K. Iga and G. Hatakoshi, "Treasure Microbox of Optoelectronics," (in Japanese) *O plus E*, vol. 38, pp. 771, 2016.

[6] M. Perales, M.-H. Yang, and J. Wu, "Low Cost Laser Power Beaming and Power over Fiber Systems," *Technical Digest of the 1st Optical Wireless and Fiber Power Transmission Conference (OWPT2019)*, OWPT-7-01, 2019.

[7] H. Helmers, E. Lopez, O. Höhn, D. Lackner, J. Schön, M. Schauerte, M. Schachtner, F. Dimroth, and A. W. Bett, "68.9% Efficient GaAs-Based Photonic Power Conversion Enabled by Photon Recycling and Optical Resonance," *Physica Status Solidi RRL*, vol. 15, pp. 2100113, 2021.

[8] S. Fafard, M. C. A. York, F. Proulx, C. E. Valdivia, M. M. Wilkins, R. Arès, V. Aimez, K. Hinzer, and D. P. Masson, "Ultrahigh Efficiencies in Vertical Epitaxial Heterostructure Architecture," *Applied Physics Letters*, vol. 108, pp. 071101, 2016.

[9] S. Fafard and D. P. Masson, "High-Efficiency and High-Power Multijunction InGaAs/InP Photovoltaic Laser Power Converters for 1470 nm," *MDPI/Photonics*, vol. 9, pp. 438, 2022.

[10] M. Miyoshi, T. Nakabayashi, K. Yamamoto, P. Dalapati, and T. Egawa, "Improved Epilayer Qualities and Electrical Characteristics for GaInN Multiple-Quantum-Well Photovoltaic Cells and Their Operation under Artificial Sunlight and Monochromatic Light Illuminations," *AIP Advances*, vol. 11, pp. 095208, 2021.

[11] J. F. Geisz, R. M. France, K. L. Schulte, M. A. Steiner, A. G. Norman, H. L. Guthrey, M. R. Young, T. Song, and T. Moriarty, "Six-Junction III-V Solar Cells with 47.1% Conversion Efficiency under 143 Suns Concentration," *Nature Energy*, vol. 5, pp. 326, 2020.

[12] K. Iga, "40 Years of Vertical-Cavity Surface-Emitting Lasers: Invention and Innovation," *Japanese Journal of Applied Physics*, vol. 57, pp. 08PA01 2018.

[13] B. D. Padullaparthi, J. Tatum, and K. Iga, "*VCSEL Industry: Communication and Sensing*," John Wiley & Sons, 2021.

[14] T. Hamaguchi, M. Tanaka, and H. Nakajima, "A Review on the Latest Progress of Visible GaN-Based VCSELs with Lateral Confinement by Curved Dielectric DBR Reflector and Boron Ion Implantation," *Japanese Journal of Applied Physics*, vol. 58, SC0806, 2019.

[15] T. Inoue, M. Yoshida, J. Gelleta, K. Izumi, K. Yoshida, K. Ishizaki, M. D. Zoysa, and S. Noda, "General Recipe to Realize Photonic-Crystal Surface-Emitting Lasers with 100-W-to-1-kW Single-Mode Operation," *Nature Communications*, vol. 13, pp. 3262, 2022.

[16] N. A. Pikhtin, S. O. Slipchenko, Z. N. Sokolova, A. L. Stankevich, D. A. Vinokurov, I. S. Tarasov, and Zh. I. Alferov, "16W Continuous-Wave Output Power from 100μm-Aperture Laser with Quantum Well Asymmetric Heterostructure," *Electronics Letters*, vol. 40, pp. 1413, 2004.

[17] P. Crump, W. Dong, M. Grimshaw, J. Wang, S. Patterson, D. Wise, M. DeFranza, S. Elim, S. Zhang, M. Bougher, J. Patterson, S. Das, J. Bell, J. Farmer, M. DeVito, and R. Martinsen. "100-W+ Diode Laser Bars Show > 71% Power Conversion from 790-nm to 1000-nm and Have Clear Route to > 85%," *Proceedings of SPIE*, vol. 6456, pp. 64560M, 2007.

[18] Y. Yamagata, Y. Kaifuchi, R. Nogawa, K. Yoshida, R. Morohashi, and M. Yamaguchi, "Highly Efficient 9xx-nm Band Single Emitter Laser Diodes Optimized for High Output Power Operation," *Proceedings of SPIE*, vol. 11262, pp. 1126203, 2020.

[19] Y. Nakatsu, Y. Nagao, T. Hirao, K. Kozuru, T. Kanazawa, S. Masui, E. Okahisa, T. Yanamoto, and S. Nagahama, "Edge-Emitting Blue Laser Diode with High CW Wall-Plug Efficiency of 50 %," *Proceedings of SPIE*, vol. 12001, pp. 1200109, 2022.

[20] T. Kageyama, K. Takaki, S. Imai, Y. Kawakita, K. Hiraiwa, N. Iwai, H. Shimizu, N. Tsukiji, and A. Kasukawa, "High Efficiency 1060nm

VCSELS for Low Power Consumption," *Technical Digest of 2009 IEEE International Conference on Indium Phosphide & Related Materials (IPRM2009)*, ThA2.1, 391–396, 2009.

[21] D. Zhou, J.-F. Seurin, G. Xu, A. Miglo, D. Li, Q. Wang, M. Sundaresh, S. Wilton, J. Matheussen, and C. Ghosh, "Progress on Vertical-Cavity Surface-Emitting Laser Arrays for Infrared Illumination Applications," *Proceedings of SPIE*, vol. 9001, pp. 90010E, 2014.

[22] Y. Wang, R. Kitahara, W. Kiyoyama, Y. Shirakura, T. Kurihara, Y. Nakanish, T. Yamamoto, M. Nakayama, S. Ikoma, and K. Shima, "8-kW Single-Stage all-Fiber Yb-Doped Fiber Laser with a BPP of 0.50 mm-mrad," *Proceedings of SPIE*, vol. 11260, pp. 1126022, 2020.

[23] Y. Katsuta and T. Miyamoto, "Design, Simulation, and Characterization of Fly-Eye Lens System for Optical Wireless Power Transmission," *Japanese Journal of Applied Physics*, vol. 58, pp. SJJE02, 2019.

[24] K. Asaba and T. Miyamoto, "Relaxation of Beam Irradiation Accuracy of Cooperative Optical Wireless Power Transmission in Terms of Fly Eye Module with Beam Confinement Mechanism," *MDPI/Photonics*, vol. 9, pp. 995, 2022.

[25] Y. Tai and T. Miyamoto, "Experimental Characterization of High Tolerance to Beam Irradiation Conditions of Light Beam Power Receiving Module for Optical Wireless Power Transmission Equipped with a Fly-Eye Lens System," *MDPI/Energies*, vol. 15, pp. 7388, 2022.

[26] J. Tang and T. Miyamoto, "Target Recognition Function and Beam Direction Control Based on Deep Learning and PID Control for Optical Wireless Power Transmission System," *Technical Digest of IEEE 9th Global Conference on Consumer Electronics (GCCE2020)*, vol. 907, 2020.

[27] A. W. S. Putra, H. Kato, and T. Maruyama, "Infrared LED Marker for Target Recognition in Indoor and Outdoor Applications of Optical Wireless Power Transmission System," *Japanese Journal of Applied Physics*, vol. 59, pp. SOOD06, 2020.

[28] K. Takahashi and T. Miyamoto, "Active Recognition of Position and Size of Solar Cell for OWPT," *Technical Digest of the 1st Optical Wireless and Fiber Power Transmission Conference (OWPT2019)*, OWPT-P-11, 2019.

[29] K. Asaba, K. Moriyama, and T. Miyamoto, "Preliminary Characterization of Robust Detection Method of Solar Cell Array for Optical Wireless Power Transmission with Differential Absorption Image Sensing," *MDPI/Photonics*, vol. 9, pp. 861, 2022.

[30] D. Tsonev, S. Videv, H. Haas, "Light Fidelity (Li-Fi): Towards all-Optical Networking," *Proceedings of SPIE*, vol. 9007, pp. 900702, 2014.

[31] K. Asaba and T. Miyamoto, "System Level Requirement Analysis of Beam Alignment and Shaping for Optical Wireless Power Transmission System by Semi–Empirical Simulation," *MDPI/Photonics*, vol. 9, pp. 452, 2022.

[32] T. J. Nugent, Jr., D. Bashford, T. Bashford, T. J. Sayles, and A. Hay, "Long-Range, Integrated, Safe Laser Power Beaming Demonstration," *Technical Digest of the 2nd Optical Wireless and Fiber Power Transmission Conference (OWPT2020)*, OWPT2-02, 2020.

[33] O. Alpert, "Long-Range Wireless Power Delivery by Infrared Light Beam – New Applications for Homes, Offices, Factories and Public Spaces," *Technical Digest of the 1st Optical Wireless and Fiber Power Transmission Conference (OWPT2019)*, OWPT-1-02, 2019.

[34] S. J. Sweeney, S. D. Jarvis, and J. Mukherjee, "Laser Power Converters for Eye-Safe Optical Power Delivery at 1550nm: Physical Characteristics and Thermal Behavior," *Technical Digest of the 1st Optical Wireless and Fiber Power Transmission Conference (OWPT2019)*, OWPT-2-04, 2019.

[35] Y. Zhou and T. Miyamoto, "200mW-Class LED Based Optical Wireless Power Transmission for Compact IoT," *Japanese Journal of Applied Physics*, vol. 58, pp. SJJC04, 2019.

[36] M. XiaoJie and T. Miyamoto, "Safety System of Optical Wireless Power Transmission by Suppressing Light Beam Irradiation to Human Using Camera," *Technical Digest of the 3rd Optical Wireless and Fiber Power Transmission Conference (OWPT2021)*, OWPT-8-03, 2021.

[37] M. XiaoJie and T. Miyamoto, "Safety System of Optical Wireless Power Transmission by Suppressing Light Beam Irradiation to Human Using Depth Camera," *Technical Digest of 25th Microoptics Conference (MOC2021)*, PO-20, 2021.

[38] K. Goto, T. Nakagawa, O. Nakamura, and S. Kawata, "An Implantable Power Supply with an Optically Rechargeable Lithium Battery," *IEEE Transactions on Biomedical Engineering*, vol. 48, pp. 830, 2001.

[39] A. Saha, S. Iqbal, M. Karmaker, S. F. Zinnat, and M. T. Ali, "A Wireless Optical Power System for Medical Implants Using Low Power Near-IR Laser," *Technical Digest of 39th Annual International Conference on IEEE Engineering in Medicine and Biology Society (EMBC)*, 1978 (2017).

[40] T. Tokuda, T. Ishizu, W. Nattakarn, M. Haruta, T. Noda, K. Sasagawa, M. Sawan, and J. Ohta, "1 mm³-Sized Optical Neural Stimulator Based on CMOS Integrated Photovoltaic Power Receiver," *AIP Advances*, vol. 8, pp. 045018, 2018.

[41] I. Haydaroglu and S. Mutlu, "Optical Power Delivery and Data Transmission in a Wireless and Batteryless Microsystem Using a Single Light Emitting Diode," *Journal of Microelectromechanical Systems*, vol. 24, pp. 155, 2015.

[42] X. Wu, I. Lee, Q. Dong, K. Yang, D. Kim, J. Wang, Y. Peng, Y. Zhang, M. Saligane, M. Yasuda, K. Kumeno, F. Ohno, S. Miyoshi, M. Kawaminami, D. Sylvester, and D. Blaauw, "A 0.04mm³ 16nW Wireless and Batteryless Sensor System with Integrated Cortex-M0+ Processor and Optical Communication for Cellular Temperature Measurement," *Technical Digest of 2018 IEEE Symposium on VLSI Circuits*, pp. 191–192, 2018.

[43] Y. Zhou and T. Miyamoto, "400 mW Class High Output Power from LED-Array Optical Wireless Power Transmission System for Compact IoT," *IEICE ELEX*, vol. 18, pp. 20200405, 2021.

[44] T. Miyamoto, K. Ueda, J. Zhang, and K. Tsuruta, "Optical Wireless Power Transmission Technology for Indoor Equipment and Mobilities," (in Japanese) *The Review of Laser Engineering*, to be published, 2023.

[45] M. Zhao and T. Miyamoto, "Optimization for Compact and High Output LED-Based Optical Wireless Power Transmission System," *MDPI/Photonics*, vol. 9, pp. 14, 2021.

[46] Y. Zhou and T. Miyamoto, "Tolerant Distance and Alignment Deviation Analysis of LED-Based Portable Optical Wireless Power Transmission System for Compact IoT," *IEEJ Transactions on Electronics, Information and Systems*, vol. 141, pp. 1274, 2021.

[47] M. Zhao and T.i Miyamoto, "1 W High Performance LED-Array Based Optical Wireless Power Transmission System for IoT Terminals," *MDPI/Photonics*, vol. 9, pp. 576, 2022.

[48] J. Tang, K. Matsunaga, and T. Miyamoto, "Numerical Analysis of Power Generation Characteristics in Beam Irradiation Control of Indoor OWPT System," *Optical Review*, vol. 27, pp. 170, 2020.

[49] Y. Liu and T. Miyamoto, "Characterization of Visible Color Filters on Optical Wireless Power Transmission System for Changing the Surface Appearance of Solar Cells," *Journal of Engineering*, vol. 2021, pp. 19, 2021.

[50] M. Ishino, T. Ohashi, T. Kitamura, A. Takamori, and K. Yamamoto, "Detection of Moving Objects and Laser Power Supply by Visible Laser Diode," (in Japanese) *The Review of Laser Engineering*, vol. 48, pp. 88, 2020.

[51] Q. Liu, M. Xiong, M. Liu, Q. Jiang, W. Fang, Y. Bai, "Charging a Smartphone over the Air: The Resonant Beam Charging Method," *arXiv*, 2105.13174v3, 2022.

[52] H. Imai, N. Watanabe, K. Chujo, H. Hayashi, and A. Yamauchi, "Beam-Tracking Technology Developed for Free-Space Optical Communication and Its Application to Optical Wireless Power Transfer," *Technical Digest of the 4th Optical Wireless and Fiber Power Transmission Conference (OWPT2022)*, OWPT5-01, 2022.

[53] K. Ueda, "Future Electric Vehicle Technology: Laser Power feeding from solar power station," (in Japanese) *Bulletin of the University of Electro-Communications*, vol. 22, pp. 63, 2010.

[54] "Contribution of Laser Technology to Carbon Neutrality in 2050," (in Japanese) *Laser Society of Japan*, 2022. www.lsj.or.jp/wp-content/uploads/Download_files/teigensho.pdf

[55] Y. Suda and T. Miyamoto, "Estimation of CO2 Emission of Dynamic Charging Electric Vehicle Using Optical Wireless Power Transmission," (in Japanese) *The Review of Laser Engineering*, to be published, 2023.

[56] N. Kawashima, K. Takeda, and K. Yabe, "Possible Utilization of the Laser Energy Transmission in Space," *Transactions of the Japan Society for Aeronautical and Space Sciences, Aerospace Technology Japan*, vol. 10, pp. Tq_1–Tq_3, 2012.

[57] N. Kawashima, K. Takeda, and K. Yabe, "Application of the Laser Energy Transmission Technology to Drive a Small Airplane," *Chinese Optics Letters*, vol. 5, pp. S109–S110, 2007.

[58] M. C. Achtelik, J. Stumpf, D. Gurdan, and K.-M. Doth, "Design of a Flexible High Performance Quadcopter Platform Breaking the MAV Endurance Record with Laser Power Beaming," *2011 IEEE/RSJ International Conference on Intelligent Robots and Systems*, 5166, 2011.

[59] Y. Kikuchi and T. Miyamoto, "Design and Characterization of Dynamic OWPT Charging to Micro-Drones," *Technical Digest of the 4th Optical Wireless and Fiber Power Transmission Conference (OWPT2022)*, OWPT5-02, 2022.

[60] K. Tsuruta and T. Miyamoto, "Investigation of Increasing of Driving Distance by Dynamic Charging Using OWPT to Small Ground-Based Mobilities," *Technical Digest of the 4th Optical Wireless and Fiber Power Transmission Conference (OWPT2022)*, OWPTp-06, 2022.

[61] M. Yamaguchi, T. Masuda, K. Araki, D. Sato, K.-H. Lee, N. Kojima, T. Takamoto, K. Okumura, A. Satou, K. Yamada, T. Nakado, Y. Zushi, M. Yamazaki, and H. Yamada, "Role of PV-Powered Vehicles in Low-Carbon Society and Some Approaches of High-Efficiency Solar Cell Modules for Cars," *Energy and Power Engineering*, vol. 12, pp. 375, 2020.

[62] F. H. Fan and S. Ishibashi, "Underwater Applications of Light Emitting Diodes," 2015 *IEEE Underwater Technology (UT)*, pp. 1–5,2015.

[63] S.-M. Kim, J. Choi, and H. Jung, "Experimental Demonstration of Underwater Optical Wireless Power Transfer using a Laser Diode," *Chinese Optics Letters*, vol. 16, pp. 080101, 2018.

[64] A. W. S. Putra, M. Tanizawa, and T. Maruyama, "Optical Wireless Power Transmission Using Si Photovoltaic through Air, Water, and Skin," *IEEE Photonics Technology Letters*, vol. 31, pp. 157, 2019.

[65] A. W. S. Putra, T. Yoshida, H. Adinanta, H. Kato, and T. Maruyama, "Optical Wireless Power Transmission through Water," *Technical Digest of the 1st Optical Wireless and Fiber Power Transmission Conference (OWPT2019)*, OWPT-8-03, 2019.

[66] T. Miyamoto and J. Li, "Research Trends in Optical Wireless Power Transmission -From Small Terminals and Mobilities to Underwater Applications-," (in Japanese) *Kogaku*, vol. 50, pp. 452, 2021.

[67] J. Li and T. Miyamoto, "Dynamic Output Power Characteristics of Optical Wireless Power Transmission from Air to Underwater," *Technical Digest of the 2nd Optical Wireless and Fiber Power Transmission Conference (OWPT2020)*, OWPT6-02, 2020.

[68] Y. Kozawa, R. Kimoto, and Y. Umeda, "Inverse Pulse Position Modulation Scheme for Underwater Visible Light Simultaneous Wireless Information and Power Transfer," *Technical Digest of the 1st Optical Wireless and Fiber Power Transmission Conference (OWPT2019)*, OWPT-8-05, 2019.

[69] S.-M. Kim and D. Kwon, "Transfer Efficiency of Underwater Optical Wireless Power Transmission Depending on the Operating Wavelength," *Current Optics and Photonics*, vol. 4, pp. 571, 2020.

[70] S. Hayashi, Y. Aoki, Y. Komuro, T. Sudo, To. Kato, W. Y. Leumg, and S. Uchida, "Laser Wireless Power Transmission in Seawater

Environment," *Technical Digest of the 3rd Optical Wireless and Fiber Power Transmission Conference (OWPT2021)*, OWPT-P-09, 2021.

[71] R. Lin, X. Liu, G. Zhou, Z. Qian, X. Cui, and P. Tian, "InGaN Micro-LED Array Enabled Advanced Underwater Wireless Optical Communication and Underwater Charging," *Advanced Optical Materials*, vol. 9, pp. 2002211, 2021.

[72] L. Tian, J. Nie, and H. Yang, "Beam Shaping for Wireless Optical Charging with Improved Efficiency," *MDPI/Crystals*, vol. 11, pp. 970, 2021.

[73] S. Hayashi, T. Kikuchi, Y. Aoki, W. Y. Leung, and S. Uchida, "Effect of the Irradiation Laser Wavelength and the Sampling Season of Seawater on Optical Wireless Power Transmission under Seawater," *Technical Digest of the 4th Optical Wireless and Fiber Power Transmission Conference (OWPT2022)*, OWPT3-03, 2022.

[74] Y. Tai, Y. Takahashi, and T. Miyamoto, "Experimental Characterization of 10m-1W Class Underwater Optical Wireless Power Transmission Using Fly-Eye Lens System," (in Japanese) *The Review of Laser Engineering*, to be published, 2023.

[75] M. Mori, H. Kagawa, and Y. Saito, "Summary of Studies on Space Solar Power Systems of Japan Aerospace Exploration Agency (JAXA)," *Acta Astronautica*, vol. 59, pp. 132, 2006.

[76] K. Takeda, M. Tanaka, S. Miura, K. Hashimoto, and N. Kawashima, "Laser Power Transmission for the Energy Supply to the Rover Exploring Ice on the bottom of the Crater in the Lunar Polar Region," *Proceedings of SPIE*, vol. 4632, 2002.

[77] M. Matsuura, "Power-Over-Fiber Using Double-Clad Fibers," *Journal of Lightwave Technology*, vol. 40, pp. 3187, 2022.

CHAPTER 6

Ultrasonic WPT

Kazuhiro Fujimori

W IRELESS POWER TRANSFER (WPT) technology has developed primarily for the purpose of transmitting and receiving electromagnetic energy. The excellent mobility of this technology has attracted
attention, and various applications are being considered, such as supplying
power not only to devices underwater or in the sea, but also to devices
inside the human body. In recent years, the development of devices that
can be placed inside the human body has been particularly active, and
various research results on these devices have been reported [1–3]. Devices
that operate inside the human body are generally designed to be powered
by on-board batteries, but devices that are implanted in the human body,
in particular, require surgical procedures that cycle through the life of the
battery, which often brings physical pain, risk, and mental anxiety to the
user. In these environments and for these purposes, an efficient and safe
WPT system using electromagnetic fields has not yet been realized. For
these reasons, the WPT system by elastic wave is more effective in some
applications. Many wireless power transfer techniques using elastic waves,
especially ultrasound, have also been studied [4,5]. In the exposure protection guidelines, ultrasound is set to have about 100 times higher power
than electromagnetic waves, and is considered to have little effect on living
organisms, as it is used for in vivo imaging such as fetal diagnosis and

DOI: 10.1201/9781003328636-6

internal organ damage. In addition, the attenuation of ultrasonic waves in the human body is smaller than that of electromagnetic waves. On the other hand, the living environment in which we live is filled with various sounds and vibrations, and these elastic waves are always present. If the energy of these elastic waves can be recovered as electrical energy, we can expect innovation in energy recycling technology as a new ubiquitous power source. Microwave WPT systems use antennas to transmit and receive electromagnetic waves. The major differences between this system and ultrasonic WPT systems can be summarized as follows: a transducer that converts electrical energy and vibration energy is used instead of an antenna, and the wavelength is small enough to be about millimeter waves. Therefore, the design and development of transducers with high conversion efficiency is essential to realize a highly efficient system. There have been several reports on wireless power transfer using ultrasonic waves, but most of these have an energy transmission efficiency of about 20%, and even the highest efficiency is only 40% [4,6–8]. This chapter explains the theory and basic device design for constructing WPT systems using ultrasonic waves, and presents examples of actual measurements using a testbed.

6.1 BASIC THEORY

This section describes the theory of elastic wave propagation, including ultrasonic waves. Like Maxwell's equations for the propagation of electromagnetic waves, the governing equations for elastic waves propagating in a fluid are the equation of continuity (6.1) and the equation of motion (6.2) shown below.

$$\frac{\partial p}{\partial t} = -\kappa \left(\frac{\partial \dot{u}_f}{\partial x} + \frac{\partial \dot{v}_f}{\partial y} + \frac{\partial \dot{w}_f}{\partial z} \right) \tag{6.1}$$

$$\frac{\partial}{\partial t} \begin{bmatrix} \dot{u}_f \\ \dot{v}f \\ \dot{w}_f \end{bmatrix} = -\frac{1}{\rho_f} \begin{bmatrix} \frac{\partial p}{\partial x} \\ \frac{\partial p}{\partial y} \\ \frac{\partial p}{\partial z} \end{bmatrix} \tag{6.2}$$

where p is the pressure increment and t is the time, respectively. κ and p represent the volume modulus and density of the fluid, respectively, and can be approximated by constants if the pressure and volume changes are assumed to be small. The subscript f indicates that the value is in fluid. u_f, v_f and w_f represent the small displacements of the fluid element in the x, y, and z directions, respectively, and their time derivatives $\partial \dot{u}_f / \partial t$, $\dfrac{\partial \dot{v}_f}{\partial t}$, and $\dfrac{\partial \dot{w}_f}{\partial t}$ correspond to the velocity change of the fluid particles due to elastic waves.

On the other hand, the governing equations for elastic waves in solids are those of Hooke's law, which describes the relationship between the strain tensor S and stress tensor T, and the equation of motion. Since both tensors are symmetric around the diagonal elements, the independent elements of the 3D stress tensor have six components. Using vectors S and T with these components, Hooke's law can be expressed as $T = [c] \cdot S$ where $[c]$ is the stiffness tensor. Matching the expression to the governing equations in fluid, the equation of motion and the time derivative of both sides of Hooke's law are expressed as follows.

$$\frac{\partial}{\partial t}\begin{bmatrix} -T_1 \\ -T_2 \\ -T_3 \\ -T_4 \\ -T_5 \\ -T_6 \end{bmatrix} = -[c] \begin{bmatrix} \dfrac{\partial}{\partial x} & 0 & 0 \\ 0 & \dfrac{\partial}{\partial y} & 0 \\ 0 & 0 & \dfrac{\partial}{\partial z} \\ 0 & \dfrac{\partial}{\partial z} & \dfrac{\partial}{\partial y} \\ \dfrac{\partial}{\partial z} & 0 & \dfrac{\partial}{\partial x} \\ \dfrac{\partial}{\partial y} & \dfrac{\partial}{\partial x} & 0 \end{bmatrix} \begin{bmatrix} \dot{u}_s \\ \dot{v}_s \\ \dot{w}_s \end{bmatrix} \qquad (6.3)$$

$$\frac{\partial}{\partial t}\begin{bmatrix} \dot{u}_s \\ \dot{v}_s \\ \dot{w}_s \end{bmatrix} = -\frac{1}{\rho_s}\begin{bmatrix} \dfrac{\partial(-T_1)}{\partial x} + \dfrac{\partial(-T_6)}{\partial y} + \dfrac{\partial(-T_5)}{\partial z} \\[2mm] \dfrac{\partial(-T_6)}{\partial x} + \dfrac{\partial(-T_2)}{\partial y} + \dfrac{\partial(-T_4)}{\partial z} \\[2mm] \dfrac{\partial(-T_5)}{\partial x} + \dfrac{\partial(-T_4)}{\partial y} + \dfrac{\partial(-T_3)}{\partial z} \end{bmatrix} \tag{6.4}$$

T_1 and T_3 represent the vertical stress in the x and z directions, respectively, and T_5 represents the shear stress. For a solid that is isotropic in three dimensions, the stiffness tensor is expressed as follows.

$$[c] = \begin{bmatrix} c_{11} & c_{13} & c_{13} & 0 & 0 & 0 \\ c_{13} & c_{11} & c_{13} & 0 & 0 & 0 \\ c_{13} & c_{13} & c_{11} & 0 & 0 & 0 \\ 0 & 0 & 0 & c_{55} & 0 & 0 \\ 0 & 0 & 0 & 0 & c_{55} & 0 \\ 0 & 0 & 0 & 0 & 0 & c_{55} \end{bmatrix} \tag{6.5}$$

Using Young's modulus E and Poisson's ratio v, which are eigenvalues of the material, each element of the stiffness tensor can be expressed by the following equations.

$$c_{11} = \frac{E(1-v)}{(1+v)(1-2v)} \tag{6.6}$$

$$c_{13} = \frac{Ev}{(1+v)(1-2v)} \tag{6.7}$$

$$c_{55} = \frac{c_{11}-c_{13}}{2} = \frac{E}{2(1-v)} \tag{6.8}$$

The governing equations in fluid should be used when discussing ultrasonic wave propagation in space or water, and the governing equations in solid when designing transducers or ultrasonic speakers to generate or receive elastic waves. At the interface between a fluid and a solid, if the boundary conditions are given so that the pressure in the fluid and the vertical stress at the surface of the solid are continuous, it is possible to obtain the physical phenomena of transmitting and receiving waves using an elastic wave device. In this case, the shear stress at the solid surface should be zero. Using Equations (6.1) and (6.2) and rearranging the equations for pressure in the x direction, for example, the following equation is obtained.

$$\frac{\partial^2 p}{\partial x^2} - \frac{\rho_f}{\kappa}\frac{\partial^2 p}{\partial t^2} = \frac{\partial^2 p}{\partial x^2} - \frac{1}{v^2}\frac{\partial^2 p}{\partial t^2} = 0 \tag{6.9}$$

This equation represents the one-dimensional wave equation for a longitudinal wave propagating in a fluid, where $v = \sqrt{\kappa/\rho_f}$ is the velocity of wave propagation and $\rho_f v$ is the intrinsic acoustic impedance determined from the volume modulus and fluid density. For example, in water, the propagation velocity depends on water temperature, water pressure, and impurity concentration, and as is well known, longitudinal waves propagate at about 1500 m/s. In this case, ultrasonic waves with a frequency exceeding 40 kHz can be considered millimeter waves with a wavelength of 37.5 mm or less. Therefore, a transducer of an easy-to-handle size is a large-aperture surface antenna.

Similar to electromagnetic waves, elastic waves emitted from an aperture large enough for the wavelength propagate as plane waves up to a certain distance and diffuse and propagate spherically beyond the limit point. This limit point is called the near-field limit and is denoted by $L = D^2/4\lambda$. Figure 6.1 illustrates this situation, where D is the diameter of the transducer. As with electromagnetic waves, the region closer to the near-field limit is called the Fresnel zone and the region farther away is called the Fraunhofer zone.

As shown in Figure 6.1, elastic waves propagate as plane waves in the Fresnel zone and spherical waves in the Fraunhofer zone.

Fresnel zone
(Plane wave)

Fraunhofer zone
(Spherical wave)

Diameter D

Transducer

Near-Field limit

$$L = \frac{D^2}{4\lambda}$$

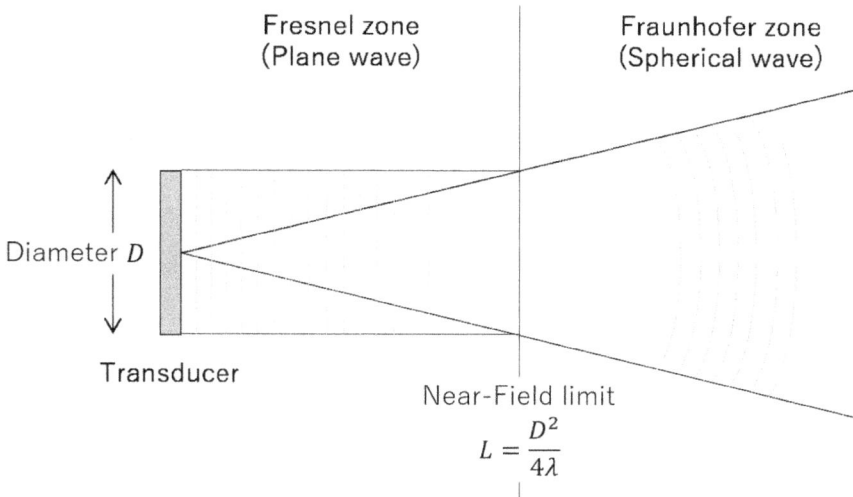

FIGURE 6.1   Elastic wave propagation in fluid.

6.2 TRANSCEIVER DEVICES

Ultrasonic speakers and transducers are commonly used to transmit and receive ultrasonic waves propagating through fluids. Both devices convert electrical energy and endo energy, and the former is suitable for transmitting and receiving large signals and the latter for small signals. This section describes the design of transducers using piezoelectric elements.

For ultrasonic wireless power transfer systems, it is essential to design and develop transducers with high conversion efficiency. Recently, multiphysics simulators have become widely available, and it is possible to obtain a solution by designing the transducer structure and the transmitter/receiver system on CAD, but the search is inefficient because it requires a large number of parameter combinations. Therefore, constraint conditions for efficiently obtaining design parameters are derived using equivalent circuits. As an example, a schematic diagram of a transducer placed in water is shown in Figure 6.2.

The <1> and <2> in the figure show the electrodes that vibrate when voltage is applied to the piezoelectric element, which has the same structure as a capacitor. The piezoelectric element between these electrodes must satisfy similar boundary conditions in <1> and <2> planes, so its thickness should be $\lambda/2$ at the design frequency. The air backing on the <1> side is designed to prevent vibrations generated by the piezoelectric element

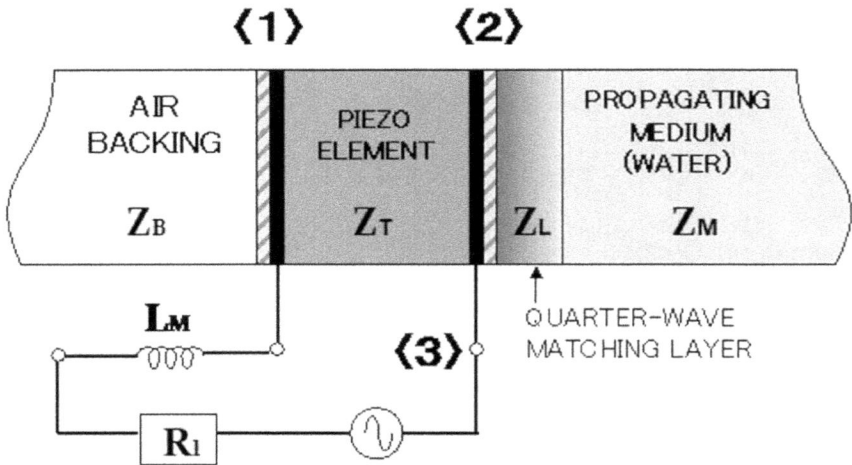

FIGURE 6.2   Physical arrangement of a piezoelectric transducer.

from propagating to the back side, and its thickness is $\lambda/4$ in the wavelength in air. On the <2> side, an acoustic matching layer of thickness $\lambda/4$, which is basically a polymer compound, is bonded to match the acoustic impedance between the vibrating surface and water. This can be considered in exactly the same way as connecting a $\lambda/4$ matching circuit in series on a distributed constant circuit. The acoustic impedance and thickness of the acoustic matching layer are Z_L and l_L, respectively. Also, Z_B, Z_T and Z_M in Figure 6.2 are the acoustic impedances of air, piezoelectric element, and water, respectively. Furthermore, L_M is the inductor used to match the piezoelectric element, which is a capacitor structure, with the power supply system, and R_l is the impedance of the power supply. Focusing on the effect of oscillation in the thickness direction of the piezoelectric element, the schematic diagram shown in Figure 6.2 can be represented by the equivalent circuit in Figure 6.3 using Mason's theory.

In this figure, the forces on the planes <1> and <2> are denoted as F_1 and F_2, and the particle velocities at the respective planes are denoted as U_1 and U_2. The voltage and current at the terminal <3>, where the power supply is connected, are shown by V and I, respectively. The left side from the transformer is the equivalent circuit of the electrical part and the right

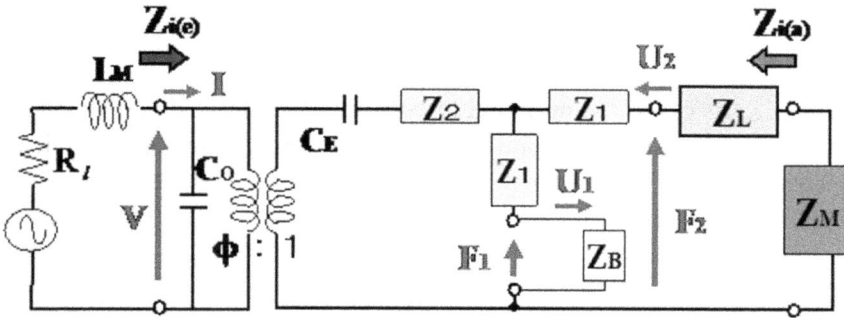

FIGURE 6.3 Equivalent circuit representation of transducer including the medium in contact.

side is the equivalent circuit of the mechanical part, where the transformer means electrical-mechanical energy conversion and ϕ corresponds to h_{33}, the material constant of the piezoelectric element. Mason's equivalent circuit allows the input impedance to be determined by mapping the force to a voltage and the particle velocity to a current. F_1, F_2 and V can be expressed by the following equations on the vibrating surfaces <1> and <2> of the piezoelectric element.

$$F_1 = -S_T Z_B U_1 \qquad (6.10)$$

$$F_2 = -Z_{L/M} U_2 \qquad (6.11)$$

$$V = -Z_{i(e)} I \qquad (6.12)$$

where S_T is the aperture area of the transducer, $Z_{L/M}$ is the synthetic acoustic impedance looking from <2> to the water side, and $Z_{i(e)}$ is the input impedance of the transducer.

First, the electrical matching conditions are derived. $Z_{L/M}$ and $Z_{i(e)}$ are expressed by the following equations.

$$Z_{L/M} = \frac{S_T Z_L^2}{Z_M} \qquad (6.13)$$

$$Z_{i(e)} = \text{Re}\{Z_{MO}\} + \text{Im}\{Z_{MO}\} + \frac{1}{j\omega C_0} \qquad (6.14)$$

Here, Z_{MO} is called the motional impedance at the terminal to which the power supply is connected, and C_0 represents the capacitance of the piezo-electric element. Vibrations in the thickness direction of the piezoelectric element occur when its thickness l_T is an integer multiple of $\lambda/2$ at the design frequency.

$$l_T = m\frac{\lambda}{2} \,(m \text{ is an integer}) \qquad (6.15)$$

In this case, Z_{MO} is expressed as follows.

$$Z_{MO} = \frac{h_{33}^2}{\omega^2 S_T} \cdot \frac{2Z_M(\cos m\pi - 1)}{(Z_L^2 + Z_M Z_B)\cos m\pi} \qquad (6.16)$$

Since the matching condition of the transducer is that the imaginary part of Equation (6.14) is equal to 0Ω and the real part is equal to the impedance R_i of the power supply, as shown in Equation (6.7), the imaginary part of Z_{MO} is 0Ω when the thickness of the piezoelectric element is as shown in Equation (6.6). Therefore, by inserting a matching inductor $L_M \,(= 1/\omega^2 C_0)$ in series with the transducer, the remaining imaginary part of Equation (6.5) can be canceled. In Equation (6.7), m must be odd because $Z_{MO} = 0\Omega$ when m is even. In this case, $Z_{i(e)}$ can be rewritten in the following simple equation.

$$Z_{i(e)} = \text{Re}\{Z_{MO}\} = \left(\frac{2h_{33}}{\omega}\right)^2 \frac{1}{Z_{L/M} + S_T Z_B} \qquad (6.17)$$

Next, the mechanical matching conditions are derived. Using Equation (6.12) and the equivalent circuit for F_2 and U_2, the acoustic impedance $Z_{i(a)}$ of the transducer can be expressed as

$$Z_{i(a)} = \frac{Z_L^2}{\dfrac{1}{S_T R_l}\left(\dfrac{2h_{33}}{\omega}\right)^2 - Z_B} \tag{6.18}$$

In Equation (6.17), the condition that $Z_{i(a)}$ must satisfy is equal to the acoustic impedance of water, Z_M. When impedance matching is performed using an acoustic matching layer without adjusting $Z_{i(a)}$, the acoustic matching layer should be selected to satisfy the following equation.

$$Z_L = \sqrt{Z_T \cdot Z_M} \tag{6.19}$$

On the other hand, by choosing parameters so that $Z_B/Z_T \ll 1$, propagation to the backward can be suppressed. As a result, the following equations, which are important in transducer design, can be obtained from Equations (6.17) and (6.18).

$$f \cdot r = \sqrt{\frac{h_{33}^2}{R_l \pi^3} \cdot \frac{1}{\dfrac{Z_L^2}{Z_M} - Z_B}} \tag{6.20}$$

where f is the design frequency and r is the aperture radius of the transducer. In general, the material constants of the piezoelectric element, acoustic matching layer, and backing material cannot be arbitrarily selected, and the transducer can be designed by determining the material to be used and the power impedance in Equation (6.20).

6.3 APPLICATIONS

In this section, based on the transducer design method described in the previous section, the characteristics of the prototype transducer and the results of ultrasonic WPT are presented as an example of applications.

Table 6.1 shows the equivalent circuit parameters of a transducer designed for a frequency of 1.2 MHz.

TABLE 6.1   Equivalent Circuit Parameters Derived from the Theoretical Equations and the Multiphysics Analysis

Parameter	Value (Unit)
h_{33} (piezoelectric deformation constant)	3.0×10^9 (V/m)
Z_T (impedance of piezoelectric element)	31.9×10^6 (kg/m²s)
Z_L (impedance of matching layer)	3.5×10^6 (kg/m²s)
Z_M (impedance of water)	1.5×10^6 (kg/m²s)
Z_B (impedance of air backing)	408 (kg/m²s)
f (design frequency)	1.2 (MHz)
r (aperture radius of piezoelectric element)	22 (mm)
l_T (thickness of piezoelectric element)	1.88 (mm)
l_L (thickness of matching layer)	0.52 (mm)
C_0 (electrostatic capacity)	9.63 (nF)
L_M (matching inductor)	1.83 μ (F)

Photographs of the prototype transducer based on the obtained design parameters and the measured input impedance are shown in Figures 6.4 and 6.5, respectively.

In Figure 6.5, the measurement results are almost consistent with the characteristics obtained by multiphysics analysis, and it can be confirmed that the real part of the impedance is 50 Ω and the imaginary part is 0 Ω at around 1.2 MHz, which is the design frequency.

Next, we present an ultrasonic WPT experiment in water using the prototype transducer as a power transmitter and the receiver device is introduced. A simplified ultrasonic WPT experimental system is shown in Figure 6.6.

The value of the load in the figure is the same as the impedance of the input source, with a 50Ω resistor connected. Transmission efficiency is defined as the power consumption at the load divided by the power delivered by the input source, and the frequency response of transmission efficiency is shown in Figure 6.7. Note that the distance between the transmitter and receiver is 50 mm.

In the measurement, a transmission efficiency of approximately 60% is confirmed around the design frequency, and the overall frequency response is in good agreement with the results of the multiphysics analysis. The main

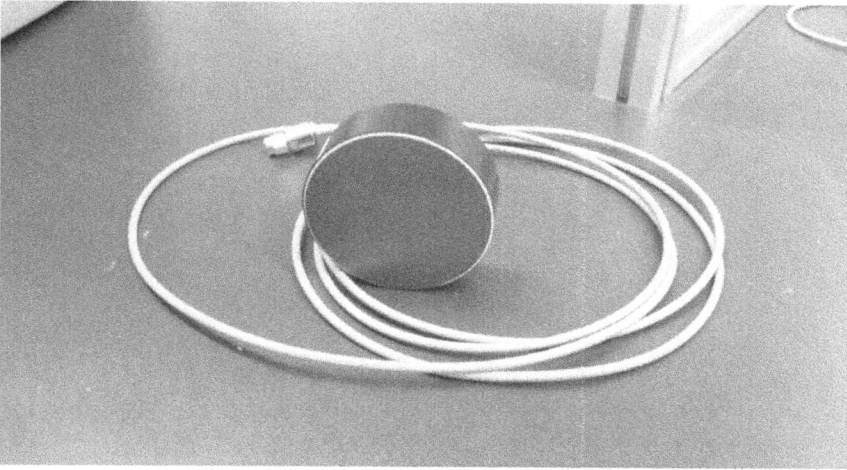

FIGURE 6.4   A prototype of the designed transducer.

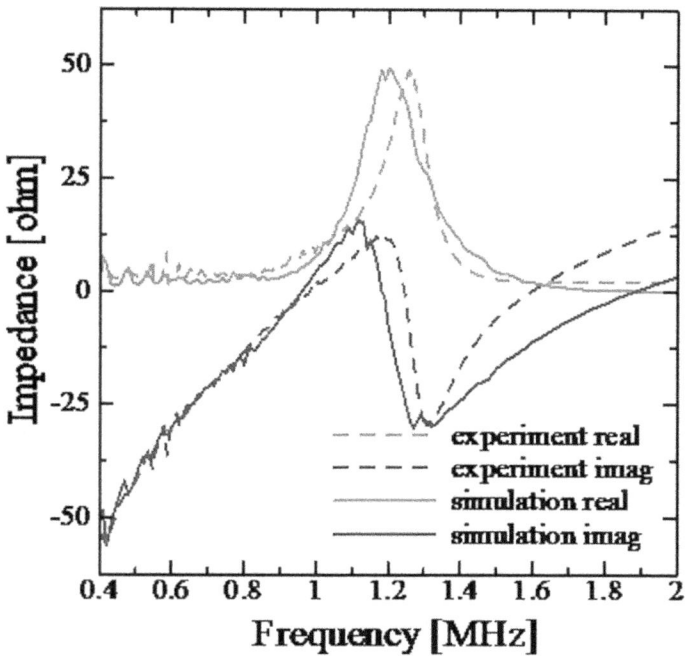

FIGURE 6.5   Frequency response of the input impedance of the prototype transducer.

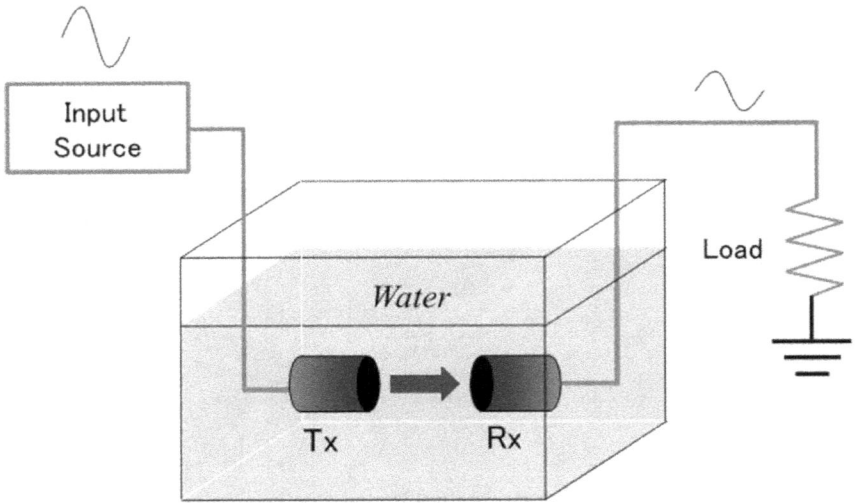

FIGURE 6.6   A simplified ultrasonic WPT experimental system.

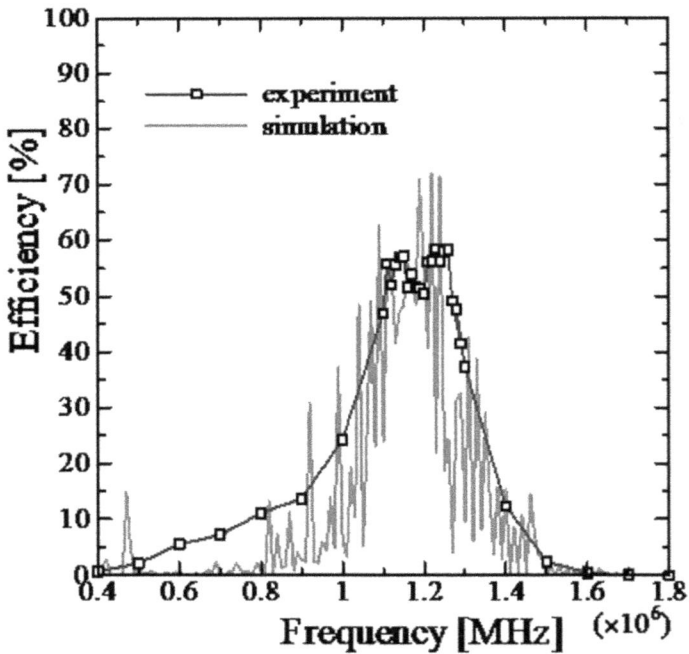

FIGURE 6.7   Frequency response of transmission efficiency in an ultrasonic WPT experiment.

reason why the maximum transmission efficiency remains at around 60% is that the electrical-to-mechanical energy conversion efficiency of the piezoelectric element is approximately 0.8–0.9, and losses occur during transmission and reception.

REFERENCES

[1] J.C. Lin, "Health Effects Space Solar-Power Stations, Wireless Power Transmissions, and Biological Implications," *IEEE Microwave Magazine*, pp. 36–42, Mar. 2002.

[2] Lin Lin, Kai-Juan Wong, Su-Lim Tan, and Soo-Jay Phee, "In-Vivo Wireless Capsule for Health Monitoring," *IEEE International Symposium on Consumer Electronics*, pp. 1–4, Apr. 2008.

[3] Kamya Yekeh Yazdandoost, and Ryuji Kohno, "Body Implanted Medical Device Communications," *IEICE Transactions on Communications*, vol. E92-B, no. 2, pp. 410–417, Feb. 2009.

[4] Takuro Yokoyama, Masaki Kyoso, Sunao Takeda, and Akihiko Uchiyama, "Development of Ultrasonic Energy Transmission System for Implanted Device," *IEICE Transactions on Fundamentals of Electronics, Communications and Computer Sciences* (in Japanese), vol. J84-A, no. 12, pp. 1565–1571, Dec. 2001.

[5] Shin-nosuke Suzuki, Shunsuke Kimura, Tamotsu Katane, Hideo Saotome, Osami Saito and Kazuhito Kobayashi, "Power and Interactive Information Transmission to Implanted Medical Device Using Ultrasonic," *Japanese Journal of Applied Physics*, vol. 41, pp. 3600–3603, Part 1, No. 5B, May 2002.

[6] F. Mazzilli, M. Peisino, R. Mitouassiwou, B. Cotte, P. Thoppay, C. Lafon, P. Favre, E. Meurville and C. Dehollain, "In-Vitro Platform to Study Ultrasound as Source for Wireless Energy Transfer and Communication for Implanted Medical Devices," *32nd Annual International Conference of the IEEE EMBS*, pp. 3751–3754, Sep. 2010.

[7] Shaul Ozeri and Doron Shmilovitz, "Ultrasonic Transcutaneous Energy Transfer for Powering Implanted Devices," *Ultrasnonics 50 Elsevier*, pp. 556–566, 2010.

[8] Alexey Denisov and Eric Yeatman, "Ultrasonic vs. Inductive Power Delivery for Miniture Biomedical Implants," *2010 International Conference on Body Sensor Networks*, pp. 84–89.

CHAPTER 7

Pros and Cons of Each WPT System

7.1 GENERAL COMPARISON

A wireless power transfer (WPT) system is classified into the near-field WPT (coupled WPT) comprising inductive and capacitive coupling WPTs, and the far-field WPT (radiative WPT) comprising radio wave, optical, and ultrasonic WPTs. Except ultrasonic WPT, all WPT systems are based on Maxwell's equations and they can be explained seamlessly. Table 7.1 presents a general comparison of the pros and cons of all the WPT systems. Several variations in the parameters can be observed, and they cannot be determined indiscriminately. However, in general, the near-field WPT with low frequency is applied as a short-distance, high-power, and high-efficiency WPT. The wireless power is transferred through electromagnetic coupling. Moreover, target detection is not necessary because the change in the position of the receiver changes the circuit parameter, resonance frequency, and impedance in the transmitter automatically. The optimization of the circuit parameters in the transmitter is necessary to ensure high transmission efficiency in the near-field WPT.

In contrast, the far-field WPT, including the ultrasonic WPT, is applied as a long-distance, low-power, and low-efficiency WPT. Moreover, it does not change the circuit transmitter despite changes in the position of the

DOI: 10.1201/9781003328636-7

TABLE 7.1 General Comparison of Pros and Cons of All Wireless Power Transfer (WPT) Systems

	Inductive coupling	Capacitive coupling	WPT via radio waves	Optical WPT	Ultrasonic
Power carrier	Magnetic field by current	Electric field by voltage	Electromagnetic waves	Light/laser (electromagnetic waves)	Sounds
Method	Electromagnetic coupling	Electromagnetic coupling	Radiation	Radiation	Radiation
Typical frequency	kHz (normal) MHz (resonance)	kHz (normal) MHz (resonance)	GHz–THz	Several hundred THz	Several hundred kHz
Method	Coil	Metal plate	Antenna	Mirror	Ultrasonic Transducer
Transmission efficiency	High	High	Low (wide beam) to high (narrow beam)	High	Middle
Target tracking/detection	Automatic recognition of changes of position	Automatic recognition of changes of position	Necessary	Necessary	Necessary
Recommendation distance	Short (<some dozen cm [normal], <several m [resonance])	Short (<some dozen cm [normal], <several m [resonance])	Long (>several m)	Long (>several m)	Long (>several m)
Typical power	High	High	Low to high	Low to high	Middle
Absorption in media	Negligible in air	Negligible in air	×Water ×Air if f is >10 GHz	×Water ×Air (rain/cloud)	×Vacuum
Safety	Magnetic field (ICNIRP) [1]	Electric field (ICNIRP) [1]	Electric field (or power) (ICNIRP) [2]	Eye-safe (ICNIRP) [3]	Thermal index and mechanical index (WUFMB)[4]
Regulation	IEC61980-2, (for EV) [5], SAE J2954 (for EV)[6], ITU-R Recommendation ITU-R SM.2110-1 [7], etc.	Under discussion	ITU-R Recommendation ITU-R SM.2151 [8], Japanese Radio Regulation (920 MHz, 2.4 GHz, 5.7 GHz) [9]	—	—

ICNIRP: International Commission on Non-ionizing Radiation Protection.
IEC: International Electrotechnical Commission.
SAE: Society of Automotive Engineers.
ITU-R: International Telecommunication Union Radiocommunication Sector.
WUFMB: World Federation for Ultrasound in Medicine and Biology.

receiver because it is an uncoupled WPT via radiation. However, target detection is essential to maintain high transmission efficiency in the far-field WPT. Although there is no theoretical limitation on the transmission of power both for the near- and far-field WPTs, there exists a technical limitation on the power, which mainly occurs when loss is converted to heat. This is a safety issue particularly in the far-field WPT because unexpected humans/living things/objects can easily enter between a transmitter and a receiver in the far-field WPT. The efficiency is dependent on the distance between the transmitter and receiver. The efficiency can be theoretically increased in the case of the far-field WPT with the use of sufficient large antennas.

In all WPT systems, the electricity must exhibit a high-frequency (>kHz) magnetic field, electric field, radio waves, laser, and ultrasonic wave, in the transmitter circuit, and it must be converted to electricity (typically direct current [DC]) in the receiver. The total efficiency of the WPT is calculated as the circuit conversion efficiency (in the transmitter) × transmission efficiency × the circuit conversion efficiency (in the receiver). With an increase in frequency, a semiconductor is used in the circuit and the efficiency of the semiconductor/circuit decreases. Thus, the total efficiency of the near-field WPT system is higher than that of far-field WPT systems because of lower frequency and shorter distance, which result in higher circuit conversion and transmission efficiencies, respectively.

The inductive and capacitive WPTs are considered as a dual circuit. The two dual quantities are magnetic and electric fields, inductance and capacitance, current and voltage, and further pairs. In the inductive WPT, the generated high-frequency magnetic field caused by a current in a coil transmits the energy wirelessly. If a higher power is required to be transmitted wirelessly, a higher current is required, which results in Ohmic loss. In contrast, in the capacitive WPT, the generated high-frequency electric field caused by a voltage in a metal plate transmits the energy wirelessly. In this, if a higher power is required to be transmitted wirelessly, a higher voltage is required. Although this does not cause much Ohmic loss, it is more likely to result in discharge. Thus, higher voltage is not suitable for semiconductors.

Recently, WPT via radio waves has been classified with wide- and narrow-beam WPTs. The wide-beam WPT is applied in a simultaneous multi-receiver system. It has small antennas, low efficiency, and low power. The wide-beam WPT is regulated worldwide. For example, new radio regulation for wide-beam WPT was established in Japan in May, 2022. In contrast, narrow-beam WPT is applied in the case of a single receiver system. It has sufficiently large antennas to realize high efficiency and power. The narrow-beam WPT has a long research and development history; however, there exists no special regulations worldwide.

For applying a far-field WPT, a narrow-beam radio wave, optical, and ultrasonic WPTs can be used. In this case, diffusion of the radio waves or the laser or the ultrasonic device must be considered. Each diffusion increases in proportion to the square of the distance from the transmitter. The far-field radio wave and optical WPTs can be compared based on Maxwell's equation and the Friis transmission formula. The diffusion of the radio waves and laser WPT is better in proportion to the square of the frequency, implying greater antenna gain for identical physical aperture of the antenna/mirror. A comparison of far-field WPT with low and high frequency, including optical WPT, is presented in Table 7.2. The diffusion and diffraction are theoretically limited; however, the circuit efficiency and its power can be improved through engineering. Further, the theoretical limitation of the transmission efficiency and antenna size can be broken by increasing the frequency though engineering. In the 1960s, 2.45 GHz was the best frequency possible for the narrow-beam WPT via radio waves in

TABLE 7.2   Comparison of the Far-field WPT with Low and High Frequency including Optical WPT

Frequency for Far Field WPT (Radio Waves and Laser)	Antenna			Propagation		Circuit	
	Transmission Efficiency (if same antenna)	Antenna Size (if same efficiency)	Beam (Phase) Control	Diffraction	Loss in Atmosphere	Circuit Efficiency	Power
Limitation by	Theory	Theory	Engineering	Theory	Physics	Engineering	Engineering
Which is better?	Higher Frequency (including OWPT)	Higher Frequency (including OWPT)	Lower Frequency	Lower Frequency	Lower Frequency	Lower Frequency	Lower Frequency

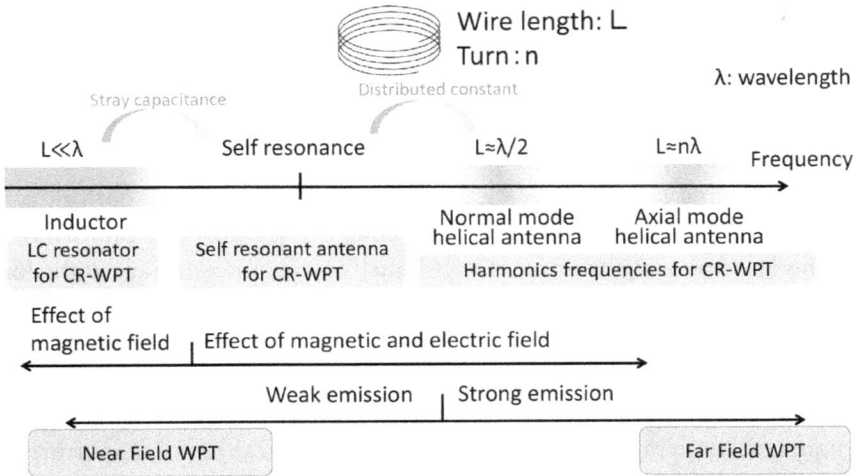

FIGURE 7.1   Relationship between near- and far-field WPTs.

Source: [2].

Brown's day. Recently, 5.7 and 24 GHz WPT systems have been produced commercially around the world, and approximately 100 GHz WPTs have been reported in studies. The trend of the frequency in the far-field WPT via radio waves is to increase the frequency.

7.2 THEORETICAL COMPARISON

As explained in the previous section, all WPT systems involving near- and far-field WPTs are based on Maxwell's equations and are explained seamlessly except ultrasonic WPT. In this section, WPT theory is explained using near-field WPT considering the inductive and radiative WPT via radio waves as representative examples. A major difference between the near- and far-field WPTs is which among the coupling of the transmitter and receiver or the radiation is the primary aspect. If a coil of the same size is applied for the WPT, it is considered as a coil for creating a low-frequency magnetic field, whose wavelength is much longer than the length of the coil in the case of the inductive WPT. Meanwhile, it is considered as a resonator with capacitance for creating a moderate-frequency magnetic field, whose wavelength is close to the length of the coil in the case of resonance WPT. Finally, it is considered as a loop antenna for creating high-frequency radio waves, whose wavelength is shorter than the length of the coil in the case of radiative WPT [10]. Figure 7.1 presents the relationship between the near- and far-field WPTs via resonance WPT [11].

In radiative WPTs, including radio wave and optical WPTs, the transmission efficiency η is calculated using the following formula obtained from Friis transmission formula.

$$P_r = \frac{\lambda^2 G_r G_t}{(4\pi r)^2} P_t = \frac{A_r A_t}{(\lambda r)^2} P_t \tag{7.1}$$

$$\eta = \frac{P_r}{P_t} = \frac{\lambda^2 G_r G_t}{(4\pi r)^2} = \frac{A_r A_t}{(\lambda r)^2} \tag{7.2}$$

where $P_r, P_t, G_r, G_t, A_r, A_t$, and λ, r are the received power, transmitted power, antenna gain of the receiving antenna, antenna gain of the transmitting antenna, aperture area of the receiving antenna, aperture area of the transmitting antenna, wavelength, and distance between the transmitting and receiving antennas, respectively. The receiving antenna whose antenna gain is G_r is placed in front of the transmitting antenna whose antenna gain is G_t. This formula can be applied only at the distance where the radio wave is assumed as a plane wave at $r \gg \frac{2D^2}{\lambda}$. Further, D_t is the diameter of the transmitting aperture antenna. In this case, receiver power and transmission efficiency are inversely proportional to the square of the distance. In case of $r < \frac{2D_t^2}{\lambda}$, the radio wave cannot be assumed as a plane wave and should be considered as a spherical wave. In this case, instead of Equation (7.2), Equations (7.3) and (7.4) are used to calculate the transmission efficiency.

$$\eta = \frac{P_r}{P_t} = 1 - e^{-\tau^2} \tag{7.3}$$

$$\tau^2 = \frac{A_t A_r}{(\lambda r)^2} \tag{7.4}$$

When the receiver is very close to the transmitter, the magnetic field should be considered to calculate the transmission efficiency instead of the radiated radio waves. The excited magnetic field H at the center at a distance of r in front of a single coil of diameter a_1 with current I is calculated using the following equation as per Biot-Savart's law.

$$H = H_r = \frac{Ia_1^2}{2\left(a_1^2 + r^2\right)^{\frac{3}{2}}} \tag{7.5}$$

Power at the center and at distance r is proportional to the square of H. A comparison with Equation (7.1) indicates its complex nature. In addition, in contrast to radiative WPT, the transmission efficiency is not calculated by the power at the receiver area. In near-field WPT, the coils are coupled electromagnetically and the efficiency is calculated based on the coupling coefficient k or mutual inductance L_m as well as an additional quality factor Q presented in Chapter 2. The mutual inductance L_m is calculated using Neumann's formula as the following equation when diameters of two single coils, a_1 and a_2, are assumed to be nearly equal and much smaller than the distance d between the coils

$$L_m \cong \mu_0 a_1 \left[log \frac{8a_1}{\left(a_1 - a_2\right)^2 + d^2} - 2 \right] [H] \tag{7.6}$$

where μ_0 is vacuum permeability. The transmission efficiency dependence of the distance r is calculated considering the combination of kQ theory and Equation (7.6); however, it is not simple. Theoretical estimation of the relationship between the distance of two coils and the transmission efficiency is shown in Chapter 2.

Furthermore, when the distance of two coils in the near-field WPT extends, unexpected radiation occurs in addition to the electromagnetic coupling (Figure 7.2). The transmission efficiency remains high, and is estimated by the coupling theory alone; however, the efficiency decreases suddenly with an increase in the distance. This is attributed to a decrease in the coupling coefficient and the unexpected radiation.

Figure 7.3 shows a trial example for comparing near-field WPT (resonance WPT) and radiative WPT at 2.45 and 5.8 GHz, at the same distance. The results of resonance WPT are experimental results and those of radiative WPT were obtained by computer simulation. The comparison revealed

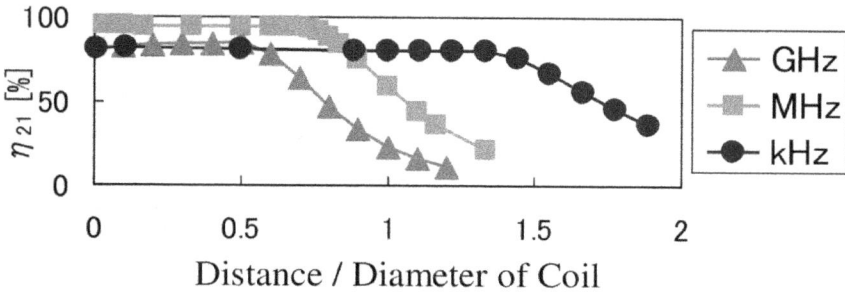

FIGURE 7.2   Relationship between transmission efficiency, ratio of distance of two coils, and diameter of coil via electromagnetic simulations.

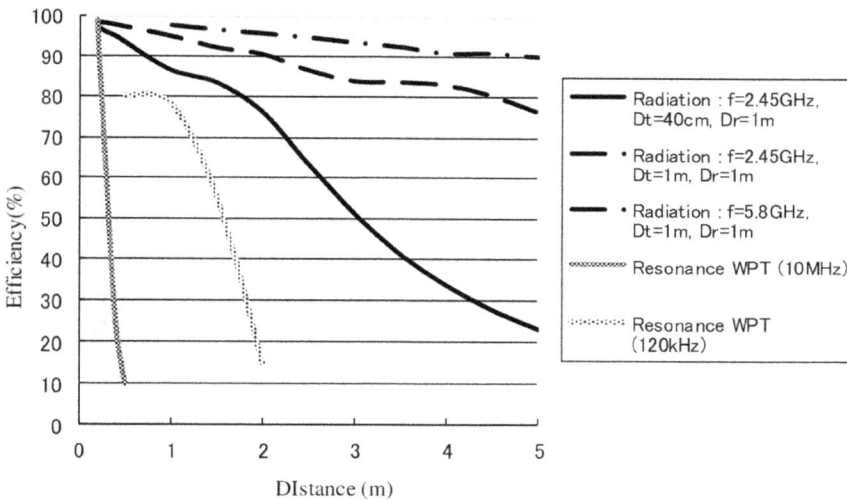

FIGURE 7.3   Comparison of transmission efficiency dependence of distance between far-field WPT via radio waves and near-field WPT (resonance WPT).

a trend of a larger decrease in the efficiency for the near-field WPT than that in the far-field WPT, even after subtracting the difference between theory and experiment; the difference was considerably large.

7.3 TECHNICAL COMPARISON

Efficiency is an important parameter for WPT. As described previously, the efficiency of the WPT is divided into transmission and circuit efficiencies. Theoretically, there is no loss factor in free space transmission in the Friis transmission formula. With sufficiently large aperture antennas, the transmission efficiency reached 100% theoretically. The coupling theory presented in Chapter 2 yielded the same result of theoretical 100% transmission efficiency. However, there is loss and scattering in real media; for example, air, glass, water, etc. There is a "radio window" of approximately 1–10 GHz in the ray path between space and ground. In addition, there is loss, reflection, and scattering in ionospheric plasmas below 1 GHz. Moreover, there is loss and scattering in the atmosphere at 10 GHz or higher, which involves optical frequency. Thus, 1–10 GHz radio waves, referred to as microwaves, are suitable for very long-distance narrow-beam WPT between space and ground. Further, water results in significant loss of the radio waves and laser. It is suitable for WPT in water by the near-field (magnetic field or electric field) and ultrasonic WPTs.

The circuit for the WPT comprises a semiconductor, inductance, capacitance, and a line with metal. In general, the loss in the semiconductor is not considered in the near-field WPT because the applied frequency is very low. The efficiency of the semiconductor in the kHz–MHz range is typically >90%. However, practically, it is better to consider the reduction of loss in the coil in the inductive WPT. To reduce the Ohmic loss in the coil, Litz wire was adopted instead of normal bold line for high power. With an increase in the frequency, the current tends to flow toward the surface, resulting in a smaller apparent cross-sectional area and greater heat generation. Litz wire is a wire fabricated by twisting multiple wires to reduce the heat generated by high frequencies in the same cross-sectional area by subdividing the conductor. Additionally, Ferrite in the coil is applied to increase the transmission efficiency in the inductive WPT, however, it is heavy and it's better not to apply it.

In the far-field WPT, the use of a semiconductor is crucial for increasing the circuit efficiency. In microwave frequency, DC-RF conversion efficiency in the high-power amplifier with a driver amplifier and matching circuits for the transmitter has already reached up to 70% at 5.75 GHz using GaN HEMT [12]. The RF-DC conversion efficiency in a rectifier with GaAs Schottky barrier diode has also reached over 90% at 2.45 and 5.8 GHz [13,14]. The world record for total efficiency at 2.45 GHz is 54% at a distance of 1.7 m in experiments [15]. In general, the efficiency and power decrease with an increase in frequency. For example, the RF-DC conversion efficiency in a rectifier at 95 GHz is 61.5% [16].

For the optical WPT, a solar cell as a receiver is an important element. As described in Chapter 5, considering an intensity 100 times that of sunlight, an efficiency of 50%–60% or more can be expected even for Si and GaAs. A monochromatic light, such as a laser, is usually applied for the OWPT and the efficiency of solar cells with respect to monochromatic light has been reported to be 40%–60% experimentally. Light-emitting diodes and semiconductor lasers are used for the transmitter in the case of the optical WPT. The efficiency of a typical commercial laser is currently 40% and is expected to be 85% in the future, as shown in Chapter 5. The total efficiency of the optical WPT is approximately 16% with typical commercial elements and approximately 52% with maximum reported value without considering air attenuation.

7.4 WPT SAFETY ISSUES

At times, users require high power for the WPT. Safety issues related to the electromagnetic field, electromagnetic waves, lasers, and ultrasonic devices are very important in developing the WPT system. A safety guideline has been discussed and established by the International Commission on Non-Ionizing Radiation Protection (ICNIRP). The safety guidance is roughly divided into time-varying electric and magnetic fields (1 Hz to 100 kHz) [1], electromagnetic fields (100 kHz to 300 GHz) [2], and laser radiation of wavelengths between 180 nm and 1,000 μm (1,666–300 THz) [3]. The guidelines are applicable to both occupational and public exposure. The difference between occupational and public exposure is based on the exposure time during working hours only or always. The safety limit for public exposure is 2–5 times lower than that for occupational exposure.

For the near-field WPT, we referred to Ref. [1], which reported that "The basis for the guidelines is two-fold. Exposure to low-frequency electric fields may yield well-defined biological responses, ranging from perception to annoyance, through surface electric-charge effects. In addition, the only well-established effects in volunteers exposed to low-frequency magnetic fields are the stimulation of central and peripheral nervous tissues and the induction in the retina of phosphenes; such as a perception of faint flickering light in the periphery of the visual field. The retina is part of the central nerve stimulation (CNS) and is regarded as an appropriate, albeit conservative, model for induced electric field effects on CNS neuronal circuitry in general." The reference levels for general public exposure to time-varying magnetic flux density B (unperturbed rms values) are shown in the following [1]:

- $4 \times 10^{-2}/f^2$ (T) @ 1–8 Hz
- $5 \times 10^{-3}/f$ (T) @ 8–25 Hz
- 2×10^{-4} (T) @ 25–400 Hz

- $8 \times 10^{-2}/f$ (T) @ 400 Hz–3 kHz
- 2.7×10^{-5} (T) @ 3 kHz–10 MHz

where f is frequency in Hz. In the frequency range above 100 kHz, RF-specific reference levels need to be considered additionally. There are other related reference levels for general public exposure to time-varying electric and magnetic field.

For the far-field WPT via radio waves, we referred to Ref. [2], which reported that "The induced electric field in the body exerts a force on both polar molecules (mainly water molecules) and free moving charged particles such as electrons and ions. In both cases a portion of the electromagnetic field energy is converted to kinetic energy, thus forcing the polar molecules to rotate and charged particles to move as a current. As the polar molecules rotate and charged particles move, they typically interact with other polar molecules and charged particles, causing the kinetic energy to be converted to heat. This heat can adversely affect health in multiple ways." The reference levels for exposure, averaged over 30 min, and whole body, to the incident power density in the range of 100 kHz to 300 GHz (unperturbed rms value) in general public circumstance are shown in the following [2]:

- 2 (W/m^2) @ >30–400 MHz
- $f_M / 200$ (W/m^2) @ 400–2000 MHz
- 10 (W/m^2) @ 2–300 GHz

where f_M is frequency in MHz. There are the other related reference levels for general public exposure to electric and magnetic fields.

The reference levels for exposure, averaged over 30 min, and whole body, to the incident power density in the range of 100 kHz to 300 GHz (unperturbed rms value) in occupational circumstance are five times larger than those for general public circumstances.

For optical WPT, we referred to Ref. [3], which reported that "Laser biological effects are the result of one or more competing biophysical interaction mechanisms: photochemical, thermal, thermo-acoustic, and optoelectric breakdown, which vary depending upon the spectral region and exposure duration. For example, in the 400Y1,400 nm band, thermal injury to the retina resulting from temperature elevation in the pigmented epithelium is the principal effect for exposure durations less than 10 s, and thermal injury to the cornea and skin occurs at wavelengths greater than 1,400 nm." The safety level required for a laser is not simple compared with the magnetic/electric/electromagnetic fields. There are certain safety standards in case of lasers, mainly to protect the eyes.

For ultrasonic WPT, reference should be made to the Thermal Index and Mechanical Index [17] for safety indices of ultrasonic devices decided by World Federation for Ultrasound in Medicine and Biology (WUFMB).

REFERENCES

[1] ICNIRP Guidelines for Limiting Exposure to Time-Varying Electric and Magnetic Fields (1 Hz – 100 kHz), *Health Physics*, vol. 99, no. 6, pp. 818–836, 2010. www.icnirp.org/cms/upload/publications/ICNI RPLFgdl.pdf

[2] ICNIRP Guidelines for Limiting Exposure to Electromagnetic Fields (100 kHz to 300 GHz), *Health Physics*, vol. 118, no. 5, pp. 483–524, 2020. www.icnirp.org/cms/upload/publications/ICNIRPrfgdl2 020.pdf

[3] ICNIRP Guidelines on Limits of Exposure to Laser Radiation of Wavelengths between 180 nm and 1,000 μm, *Health Physics*, vol. 105, no. 3, pp. 271–295, 2013. www.icnirp.org/cms/upload/publications/ ICNIRPLaser180gdl_2013_2020.pdf

[4] S. B Barnett 1, G. R Ter Haar, M. C Ziskin, H. D Rott, F. A Duck, K. Maeda, "International Recommendations and Guidelines for the Safe Use of Diagnostic Ultrasound in Medicine", *Ultrasound in Medicine and Biology*, vol. 26, pp. 335–366, 2000.

[5] IEC TS 61980-2:2019, "Electric Vehicle Wireless Power Transfer (WPT) Systems – Part 2: Specific Requirements for Communication Between Electric Road Vehicle (EV) and Infrastructure", 2019. https://webstore.iec.ch/publication/31050

[6] SAE Journal, "Wireless Power Transfer for Light-Duty Plug-in/ Electric Vehicles and Alignment Methodology", vol. 2954, 2020. www.sae.org/standards/content/j2954_202010/

[7] ITU-R Recommendation ITU-R SM.2110-1, "Frequency Ranges for Operation of Non-beam Wireless Power Transmission (WPT) Systems", 2017 and 2019. www.itu.int/rec/R-REC-SM.2110/en

[8] ITU-R Recommendation ITU-R SM.2151-0, "Guidance on Frequency Ranges for Operation of Wireless Power Transmission via Radio Frequency Beam for Mobile/Portable Devices and Sensor

Networks", 2022. www.itu.int/dms_pubrec/itu-r/rec/sm/R-REC-SM.2151-0-202209-I!!PDF-E.pdf

[9] N. Shinohara, "History and Innovation of Wireless Power Transfer via Microwave", *IEEE Journal of Microwaves*, vol. 1, no. 1, pp. 218–228, 2021.

[10] H. Hirayama, "Chapter 3 Basic Theory of Resonance Coupling WPT", in "*Wireless Power Transfer: Theory, Technology, and Applications*" edited by Naoki Shinohara, Inst of Engineering & Technology, pp. 23–36, 2018.

[11] H. Hirayama, T. Amano, N. Kikuma, and K. Sakakibara, "An Investigation on Self-Resonant and Capacitor-Loaded Helical Antennas for Coupled-Resonant Wireless Power Transfer", *IEICE Transactions on Communications*, vol. E96b, no. 10, pp. 2431–2439, 2013.

[12] R. Ishikawa, "A High-Gain and High-Efficiency Amplifier Module for DC-RF Power Conversion in Wireless Power Transfer System", *Proceedings of the URSI AT-RASC2022*, Mo-WS2-PM2-1, 2022.

[13] C. Wang, B. Yang, and N. Shinohara, "Study and Design of a 2.45GHz Rectifier Achieving 91% Efficiency at 5-W Input Power", *IEEE Microwave, and Wireless Components Letters*, vol. 31, no. 1, pp. 76–79, 2021.

[14] N. Sakai, K. Noguchi, and K. Itoh, "A 5.8-GHz Band Highly Efficient 1-W Rectenna with Short-Stub-Connected High-Impedance Dipole Antenna", *IEEE MTTS Transactions*, vol. 69, no. 7, pp. 3558–3566, 2021.

[15] W. C. Brown, "The History of Power Transmission by Radio Waves", *IEEE Transactions on MTT*, vol. 32, no. 9, pp. 1230–1242, 1984.

[16] H. Kazemi, "61.5% Efficiency and 3.6 kW/m² Power Handling Rectenna Circuit Demonstration for Radiative Millimeter Wave Wireless Power Transmission", *IEEE MTTS Transactions*, vol. 70, no. 1, pp. 650–658, 2022.

[17] S. B Barnett 1, G. R Ter Haar, M. C Ziskin, H. D Rott, F. A Duck, K. Maeda, "International Recommendations and Guidelines for the Safe Use of Diagnostic Ultrasound in Medicine", *Ultrasound in Medicine and Biology*, vol. 26, pp. 335–366, 2000.

Index

For Product Safety Concerns and Information please contact our EU
representative GPSR@taylorandfrancis.com
Taylor & Francis Verlag GmbH, Kaufingerstraße 24, 80331 München, Germany